D1309494

# ELECTRICAL NETWORK SCIENCE

**ROBERT B. KERR**

*Duke University*

PRENTICE-HALL, INC. Englewood Cliffs, New Jersey   07632

*Library of Congress Cataloging in Publication Data*

KERR, ROBERT BLACKBURN, 1929–
  Electrical network science.

  1. Electric networks.  2. Electric circuits.
3. Electric engineering—Mathematics.  I. Title.
TK454.2.K47      621.319'2      76–45358
ISBN 0–13–247627–4

10  9  8  7  6  5  4  3  2  1

Printed in the United States of America

PRENTICE-HALL INTERNATIONAL, INC., *London*
PRENTICE-HALL OF AUSTRALIA, PTY. LTD., *Sydney*
PRENTICE-HALL OF CANADA, LTD., *Toronto*
PRENTICE-HALL OF INDIA PRIVATE LIMITED, *New Delhi*
PRENTICE-HALL OF JAPAN, INC., *Tokyo*
PRENTICE-HALL OF SOUTHEAST ASIA PTE. LTD., *Singapore*
WHITEHALL BOOKS LIMITED, *Wellington, New Zealand*

# Contents

073119

# *Preface*

   The purpose of this text is to introduce the reader to some of the most basic and important concepts of linear electrical networks, in particular, and to linear systems described by differential or difference equations, in general. It is assumed that the reader has some basic familiarity with both differential and integral calculus, at a level equivalent to the usual college freshman introductory courses in these subjects, and a physics background comparable to that usually obtained in freshman physics courses. A prior background in differential or difference equations is not assumed.

   In the first two chapters the physical principles underlying linear network models are discussed in more detail than is found in most networks texts. These two chapters are intended as a basic introduction to electrical engineering. The reader who feels comfortable with the physical laws and concepts in these chapters has a good foundation for further study in the areas of circuits, fields, and solid-state theory. For this reason these two chapters contain brief qualitative discussions of certain topics (e.g., conduction mechanisms in conductors and semiconductors, junction diodes) not usually found in linear circuits texts, as well as a basic detailed treatment of the usual circuit variables and parameters, starting from Coulomb's law, the Biot–Savart law, and Faraday's law. Fortunately, a complete treatment of field theory is not necessary in such an introduction to networks and systems; the $\bar{D}$ and $\bar{H}$ vectors are not introduced in these chapters, and no use is made of the differential field operators.

   Chapters 3 through 9 contain the primary material of the text, topics usually considered indispensable in preparing electrical engineering students (in particular) for more advanced junior- and senior-level courses. The goal has been to present concepts in as logical a sequence as possible. The solution of the network equations begins with simple initial condition and step-function responses, followed by the steady-state response to sinusoidal forcing functions. The reader is thus introduced early to frequency-response ideas (Bode and Nyquist plots), to which most students seem to relate easily. Following this, the Fourier series and Fourier integral discussions form a very logical progression. In contrast to the sequences in certain other books, this text postpones the treatment of Laplace transforms and complex frequency (Chapter 8) until after the reader is thoroughly familiar with real frequency analysis; when Laplace transforms and the complex frequency plane do appear, the student has better insight into their basic nature. This approach also tends to emphasize the importance of Fourier methods and spectral analysis. Laplace transforms are not deemphasized but they are put into a more proper perspective.

An entire chapter (Chapter 7) is devoted to impulse functions, impulse response, and convolution, with emphasis on discrete-time simulation of differential systems by numerical convolution. A useful formula for numerical convolution using straight-line interpolations between signal sample values is derived, and a digital computer simulation exercise is given as a problem. This topic is taken up again in the portions of Chapter 9 that deal with discrete-time systems.

Chapter 9 introduces the state-variable analysis of both continuous- and discrete-time circuits and systems. The treatment here is more general and somewhat more set apart than that found in most circuits texts, where the discussion is usually limited to state-space models of electrical circuits only. It is hoped that this chapter will make clear to the reader the generality of the state-variable approach and provide a transition to more general systems courses.

The reader, in attempting to solve the problems at the end of each chapter, should find Appendix C helpful. In this appendix detailed solutions are given for problems that are related (but not identical) to certain selected problems in the text proper. These worked examples are numbered to correspond to the text problems to which they are related, and are an addition to the examples that appear throughout the text itself.

The author wishes to thank Dr. Harry A. Owen, Jr., and Dr. Herbert Hacker, Jr., who, in their capacities as chairmen of the Electrical Engineering Department of Duke University, have supported the preparation of the manuscript of this text, and also Dr. Paul P. Wang, who has offered much constructive criticism. Special thanks are also due Mr. Steven Sharpe, who is primarily responsible for the preparation of the Solutions Manual and who has contributed many helpful suggestions and corrections to the text. Finally, the author expresses his appreciation to the students, too numerous to name here, who have contributed suggestions and corrections in the course of studying this text material in the form of teaching notes.

*Durham, North Carolina*                                          ROBERT B. KERR

# ELECTRICAL
# NETWORK SCIENCE

# 1

# Electric Fields
# and Related Phenomena

In this chapter and in Chapter 2 we shall be concerned with some of the physical concepts necessary for a good understanding of electrical phenomena. Of course, many modern electrical engineers do not have any direct involvement with "electrical physics" in the course of their work. For example, many engineers involved in the analysis or design of control systems, communication systems, computers, and many other types of systems deal almost exclusively with mathematical models of signals, circuits, and devices, without much direct concern for the origin and accuracy of those models. Ultimately, of course, one must worry about how well the mathematical model represents the physical world; otherwise, the analysis and design are simply exercises. Hence, although the greater part of this book is involved with manipulating mathematical models of signals, circuits, and systems, the first two chapters are devoted to a discussion of the origin and accuracy of certain mathematical models of general importance in electrical engineering.

In our present-day technology, the range and diversity of devices that involve electricity and magnetism are enormous; it may not be evident to the layman that a large power generator, a high-voltage transmission line, a radar antenna, a sophisticated electronic control system, and a microelectronic circuit involving hundreds of transistors and diodes have anything in common. The mathematical models used to describe all these devices will, however, involve certain common *variables* (e.g., voltage, current, charge, flux linkages) and certain *parameters* (e.g., resistance, inductance, capacitance, mutual inductance); our introduction to electrical engineering thus most logically begins with a discussion of the definitions and relationships of these (and other) important variables and parameters. These definitions and relationships, in turn, can be thoroughly understood only by considering certain aspects of electric and magnetic fields and certain mathematical operations involving field quantities. Fortunately, a complete treatment of electric and magnetic fields is not essential for this purpose; hence we shall be highly selective and discuss only those concepts which are essential to a good understanding of the important circuit variables and parameters. Some of these field ideas may already be familiar to the reader; if so, the following sections may be considered a review of certain salient aspects of the subject.

## 1.1 FIELDS: SOME MATHEMATICAL PRELIMINARIES

### Scalars and Vectors: Fields

The reader has undoubtedly had some prior experience with scalar and vector concepts in the resolution of forces and velocities. A *vector* (mathematically, a three-dimensional Euclidean vector) is a quantity that has associated with it both a magnitude (length) and direction; a *scalar* is characterized by a magnitude only. A force is thus a vector quantity; temperature and mass density, for example, are scalars. In this book vector quantities will be distinguished by a line placed above a symbol.

It will be assumed that the reader is already familiar with the resolution of vectors into scalar components, at least in rectangular coordinates, and with the addition and subtraction of vectors by adding or subtracting their corresponding components. Products of vector quantities, however, may not be familiar to the reader at this point, and these will be discussed shortly.

If, at each point in some region of space, a vector is defined, then a *vector field* is said to exist in that region of space. In a pipe carrying fluid, for example, at any instant of time one may define a velocity vector at every point interior to the pipe; this constitutes a vector (velocity) field. At each point inside the pipe, one might also define a temperature or density; these would constitute *scalar fields*. In a uniform scalar field, the scalar has the same value at all points in the field. In a uniform vector field, the vector has the same magnitude and same direction at all points in the field. For example, the vector field denoted by

$$\bar{F} = A\bar{i}_x, \tag{1.1}$$

where $A$ is a constant (not dependent on $x$, $y$, or $z$) and $\bar{i}_x$ is a unit vector (i.e., a vector of unit length and direction indicated by the subscript), is a uniform vector field, as indicated in Figure 1.1. In a general vector field, of course, each scalar component may be some function of each of the coordinates; we may indicate this by

$$\bar{F} = A_x(x, y, z)\bar{i}_x + A_y(x, y, z)\bar{i}_y + A_z(x, y, z)\bar{i}_z. \tag{1.2}$$

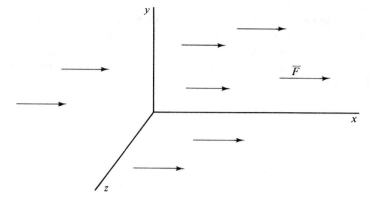

**Figure 1.1**  Uniform Field in the *x* Direction

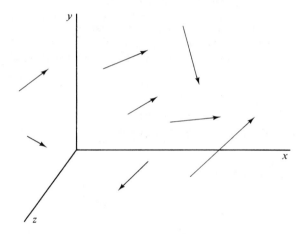

**Figure 1.2**  General Vector Field

Such a field might appear as in Figure 1.2. The scalar components of the field may also be functions of time. If they are, the field is said to be a time-varying one; if not, the field is called static or time-invariant.

### Scalar and Vector Products

There are two useful ways of defining the product of two vector quantities; the first has a scalar as its result and is called a *scalar* (or *dot*) *product*. The scalar product of two vectors $\bar{A}$ and $\bar{B}$ is denoted by $\bar{A} \cdot \bar{B}$ and may be defined by the relationship

$$\bar{A} \cdot \bar{B} \triangleq AB \cos \theta, \tag{1.3}$$

where $(A \cos \theta)$ may be interpreted as the projection of $\bar{A}$ in the direction of $\bar{B}$, as indicated in Figure 1.3. If, for example, a particle subject to a force $\bar{A}$ moves a distance indicated by the vector $\bar{B}$, then the work done by the force on that particle is $\bar{A} \cdot \bar{B}$. Note that the dot product *commutes* (i.e., $\bar{A} \cdot \bar{B} = \bar{B} \cdot \bar{A}$). If $\bar{A}$ and $\bar{B}$ are parallel, then $\bar{A} \cdot \bar{B} = AB$; if they are perpendicular, then $\bar{A} \cdot \bar{B} = 0$.

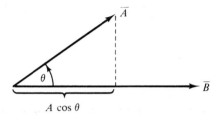

**Figure 1.3**  Scalar Product

The second type of product results in a vector quantity and is called a *vector* (or *cross*) *product*, denoted by $\bar{A} \times \bar{B}$. This product, being a vector, is slightly more complicated than the scalar product and may be defined as

$$\bar{A} \times \bar{B} \triangleq AB \sin \theta \, i_\perp, \tag{1.4}$$

where $i_\perp$ is defined to be a unit vector perpendicular to both $\bar{A}$ and $\bar{B}$, and with a sense given by a right-hand rule (Figure 1.4). By this is meant that it

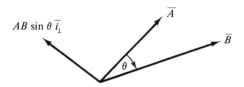

**Figure 1.4**  Vector Product

points in the direction in which a right-handed screw would advance if rotated from $\bar{A}$ to $\bar{B}$ through the smallest angle. From this definition it follows that $\bar{B} \times \bar{A}$ is simply the negative of $\bar{A} \times \bar{B}$, and if $\bar{A}$ and $\bar{B}$ are parallel, $\bar{A} \times \bar{B} = \bar{B} \times \bar{A} = 0$. If $\bar{A}$ and $\bar{B}$ are perpendicular, then $\bar{A} \times \bar{B} = ABi_\perp$. It should be carefully noted that the cross product is not commutative, because of the opposite sense of $\bar{A} \times \bar{B}$ and $\bar{B} \times \bar{A}$.

Although the definitions given here for the scalar and vector products are independent of the particular coordinate system used, expressions may of course be derived for these products in terms of the vector coordinates in any particular system. In the development to follow, we will not make use of these particular expressions, but we will give (without derivation) the expressions in rectangular coordinates. In a rectangular system of coordinates,

$$\bar{A}\cdot\bar{B} = A_x B_x + A_y B_y + A_z B_z \tag{1.5}$$

and

$$\bar{A}\times\bar{B} = \begin{vmatrix} \bar{i}_x & \bar{i}_y & \bar{i}_z \\ A_x & A_y & A_z \\ B_x & B_y & B_z \end{vmatrix}. \tag{1.6}$$

[Vertical bars ($\|$) indicate a determinant.]

### *Line Integrals*

A line integral is a scalar quantity that is associated with any particular *path* in a vector field. Consider a path from point *a* to point *b* in a vector field, $\bar{V}(x, y, z)$, as indicated in Figure 1.5. To each point along the path, the vector field associates a magnitude and direction. Suppose that the path is broken

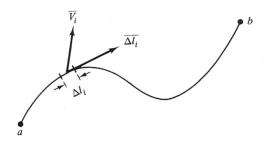

**Figure 1.5**  Line Integral as a Limit

up into small increments of length $\Delta l$. To each increment, $\Delta l_i$, we may associate a vector $\overline{\Delta l_i}$ having a length $\Delta l$ and a direction tangent to the path in the direction of travel along the path. If the vector field $\bar{V}$ is continuous in its spatial coordinates, we may assume that, in the limit, $\Delta l$ is so small that over each little increment the vector field is constant. If $\bar{V}_i$ is the value of the vector field at the increment $\Delta l_i$, we form the product of the component of $\bar{V}_i$ in the direction of travel along the path and the incremental path length $\Delta l_i$ by taking the scalar product, $\bar{V}_i \cdot \overline{\Delta l_i}$. The line integral may then be defined as the limiting value of the sum of these scalar products as $\Delta l \to 0$; i.e.,

$$\lim_{\substack{\Delta l \to 0 \\ N \to \infty}} \sum_{i=1}^{N} \bar{V}_i \cdot \overline{\Delta l_i} \triangleq \int_a^b \bar{V}\cdot\overline{dl}. \tag{1.7}$$

Thus, roughly speaking, the line integral involves adding up the component of the vector field in the direction of travel, multiplied by the incremental path length, from one point on the path to another.

Probably the most common physical interpretation of the line integral is that of the work done in moving an object in a force field. If $\bar{F}$ represents a force field, then $\int_a^b \bar{F} \cdot \overline{dl}$ is the work done by the field in moving an object along the path from point $a$ to point $b$. Conversely, $-\int_a^b \bar{F} \cdot \overline{dl}$ represents the work done by an external agent in moving the object from $a$ to $b$. Thus if $\int_a^b \bar{F} \cdot \overline{dl}$ is a positive quantity, the field supplies energy to the external agent, and vice versa if the line integral is negative.

In certain fields, the line integral will depend only on the end points of the path, not on the particular path chosen to join those end points. In this case, for any pair of points $a$ and $b$, the line integral $\int_a^b \bar{V} \cdot \overline{dl} = -\int_b^a \bar{V} \cdot \overline{dl}$ will be independent of the path chosen, and it follows that the line integral around any closed contour (as illustrated in Figure 1.6) will be zero. This is indicated by writing

$$\oint \bar{V} \cdot \overline{dl} = 0, \tag{1.8}$$

and such a field is said to be *conservative*.

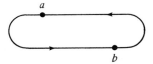

**Figure 1.6**   Line Integral Around a Closed Contour

### Example 1.1

Consider a two-dimensional radial field, given by

$$\bar{V} = \frac{1}{r^2} \bar{i}_r, \tag{1.9}$$

where $\bar{i}_r$ is a unit vector in the radial direction, as indicated in Figure 1.7. Consider also the contour as indicated in the figure. The integral around this contour may be expressed as the sum of four parts,

$$I = \int_a^b + \int_b^c + \int_c^d + \int_d^a. \tag{1.10}$$

The integrals from $b$ to $c$ and from $d$ to $a$ will be zero, since over these segments the path is everywhere perpendicular to the field, and hence the scalar product

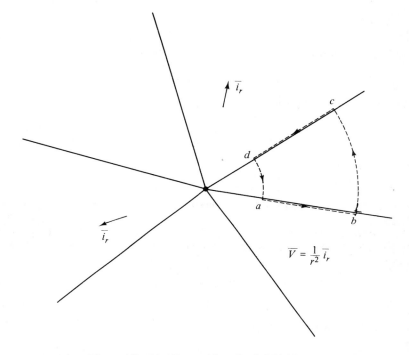

**Figure 1.7**  Line Integral in a Radial Field

involved in the definition of the line integral causes these contributions to be zero. In going from $a$ to $b$ and $c$ to $d$ we travel parallel and antiparallel to the field, respectively, so the complete contour integral becomes

$$I = \oint \bar{V} \cdot \overline{dl} = \int_a^b \frac{dr}{r^2} + 0 + \int_d^c \left(-\frac{dr}{r^2}\right) + 0 = 0. \qquad (1.11)$$

In this case the chosen contour was particularly easy to integrate around, but the same result may be shown to hold for any closed contour in this field; i.e., the field is conservative.

*Example 1.2*

Let us form the line integral from the origin to the point $(1, 1)$ in the two-dimensional uniform field in Figure 1.8. First, consider the path composed of the two straight-line segments $(0, 0)$ to $(1, 0)$ and $(1, 0)$ to $(1, 1)$. Along this path

$$I = \int_0^1 A\bar{i}_x \cdot \overline{dx} + \int_0^1 A\bar{i}_x \cdot \overline{dy}$$

$$= \int_0^1 A \, dx + 0 = A. \qquad (1.12)$$

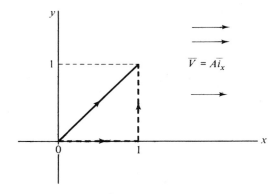

**Figure 1.8** Line Integral in a Uniform Field

Now consider integrating along the straight line connecting the origin and the point (1, 1). Along this radial line, $\overline{dl} = \overline{dr}$, and we may write

$$I = \int_0^{\sqrt{2}} \overline{V} \cdot \overline{dr} = \int_0^{\sqrt{2}} A\overline{i}_x \cdot \overline{dr}$$

$$= \int_0^{\sqrt{2}} A \cos \theta \, dr = \frac{A}{\sqrt{2}} \int_0^{\sqrt{2}} dr = A. \qquad (1.13)$$

The same result will hold for any closed contour in this field, which is again conservative. An example of a nonconservative field is given in Problem 1.5.

### Surface Integrals

Consider a surface (not necessarily a plane surface) in a vector field, as indicated in Figure 1.9. How much of the vector field cuts or flows through the surface? One has an intuitive feel for what these terms might mean, but we may be precise by defining a quantity called a *surface integral*. Suppose that the surface is divided into small incremental components of area $\Delta S$.

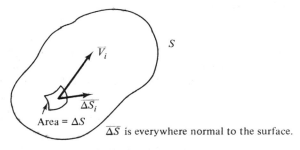

$\overline{\Delta S}$ is everywhere normal to the surface.

**Figure 1.9** Surface Integral as a Limit

Let $\overline{\Delta S_i}$ represent a vector of magnitude $\Delta S$ perpendicular to the surface at the increment $\Delta S_i$. If the surface is continuous at the point in question, we may in the limit consider $\Delta S_i$ to be so small that it is represented by a plane, and if the field is continuous, it may be considered constant over this increment of surface. We form the scalar produce of $\overline{V}_i$ and $\overline{\Delta S_i}$ and sum up the contributions of each increment by integration:

$$\lim_{\substack{N\to\infty \\ \Delta S\to 0}} \sum_{i=1}^{N} \overline{V}_i \cdot \overline{\Delta S_i} \triangleq \int_S \overline{V}\cdot\overline{ds}. \tag{1.14}$$

Hence, roughly speaking, we add up the component of the field normal to the surface multiplied by the incremental surface area, over the entire surface; this integral is called the *flux* of the vector field over (or through) the surface $S$.

As a simple mechanical illustration of the surface integral, suppose that the vector field is the velocity field of a fluid in a pipe or stream. If a surface is chosen so that all the fluid flows through the surface, the surface integral of this velocity field is simply the volume rate of flow of the fluid. (Dimensionally, velocity $\times$ area = volume per unit time.)

### Example 1.3

Consider the surface of a unit cube in a uniform field, as shown in Figure 1.10. Let us compute the flux *out* of this cube (we therefore use the outward normal to the surface at all points). Over four of the sides of the cube the surface integral is zero, since $\overline{V}$ and $\overline{dS}$ are perpendicular. Over the other two sides we have

$$I = \int_0^1 \int_0^1 A\, dx\, dz + \int_0^1 \int_0^1 A(-dx\, dz) = 0. \tag{1.15}$$

Thus the net outward (or inward) flux is zero. In fact, the net flux through

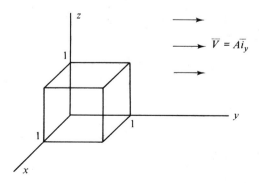

**Figure 1.10** Surface Integral in a Uniform Field

any closed surface in this field is zero; a field having this property is called *solenoidal*.

## 1.2 THE ELECTRIC FIELD: VOLTAGE

### Charge and Coulomb's Law

In the equilibrium state, the atoms of a material are uncharged; i.e., in the classical atomic model, there are as many negatively charged particles (electrons) in orbit around the nucleus as there are positively charged particles (protons) in the nucleus, resulting in a net charge of zero. Depending upon the type of atom (i.e., whether or not it is metallic, its valence, etc.) and the way that it is bonded to its neighbors, the negative charge may be either very loosely or very tightly bound to the nucleus. In any case, it is possible by various mechanical, electrical, or chemical means to separate electrons from the nucleus of the atom to form particles that have a net charge (ions), which will then exert a force upon one another. In the famous oil-drop experiment, Millikan showed that it is not possible to continuously vary the amount of charge carried by a body; rather, the charge always appears as an integer multiple of a certain smallest amount (or quantum) of charge, which is then taken to be the magnitude of the charge carried by a single electron or proton. The practical engineering unit of charge is the coulomb (C), which is equal to $6.24 \times 10^{18}$ charge quanta, thus making the charge on a single electron $-1.602 \times 10^{-19}$ C. This unit was derived historically from the ampere, a unit of electrical current, and unfortunately represents an enormous amount of charge (see Problems 1.8 and 1.9); more usual amounts of charge are measured in units of $10^{-6}$ or $10^{-12}$ C.

The most basic force relationship involving separated charges is *Coulomb's law*, which states that the force between two point charges (i.e., charges concentrated in sufficiently small volumes that these volumes may be treated as points) varies directly as the product of the two charges and inversely as the square of the distance by which they are separated; this force acts along the line determined by the two points. This law may be expressed in vector form (Figure 1.11) by writing

$$\bar{F}_{21} = (\text{const.}) \frac{q_1 q_2}{r^2} \bar{i}_{21} \tag{1.16}$$

**Figure 1.11**  Coulomb's Law

where $q_1$ and $q_2$ are the two charges (in coulombs), $r$ is the distance (meters), $\bar{\imath}_{21}$ is a unit vector pointing from charge $q_1$ toward charge $q_2$, and $\bar{F}_{21}$ is the force (newtons) exerted on charge $q_2$ by the presence of $q_1$. The proportionality constant is found experimentally to be (MKS units) $8.98740 \times 10^9$, or approximately $9 \times 10^9$, if the charges are in free space.

Equation 1.16 may also be written in the form

$$\bar{F}_{21} = \frac{q_1 q_2}{4\pi\epsilon r^3} \bar{r}_{21}, \tag{1.17}$$

where $\bar{r}_{21}$ is defined to be the *radius vector* from $q_1$ to $q_2$. In this form we need an extra $r$ in the denominator, since the magnitude of $\bar{r}_{21}$ is $r$ rather than unity. The constant is now written as $1/4\pi\epsilon$; $\epsilon$ is called the *permittivity* of the medium. When the $4\pi$ is written explicitly in this manner, the value of $\epsilon$ is said to be in *rationalized units* (writing the $4\pi$ explicitly in Coulomb's law guarantees that it will *not* appear in many other formulas, and thus simplifies much theory derived from Coulomb's law). To express the force exerted on charge $q_1$ by the presence of $q_2$, we need only reverse the order of the subscripts in equation 1.17. It should be emphasized, however, that the force given by Coulomb's law will be correct only if both charges are stationary with respect to the observer; if they are not, an additional force will be present which will be discussed later.

### The Electrostatic Field

Forces between charges are most easily analyzed by postulating the existence of a vector field, called the *electric field* or *electrical field intensity*. Consider a single point charge, $q_1$, as in Figure 1.12. If we place a second

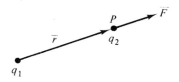

**Figure 1.12**  Pertinent to Definition of Electric Field

point charge at point $P$, having a radius vector $\bar{r}$ from $q_1$, we may say that the force $\bar{F}$ exerted on the second charge is

$$\bar{F} = q_2 \bar{E}, \tag{1.18}$$

where $\bar{E}$ is the electric field intensity at point $P$ due to $q_1$ and is given by

$$\bar{E}_{(\text{due to } q_1)} = \frac{q_1}{4\pi\epsilon r^3} \bar{r} \quad (\text{newtons/coulomb}). \tag{1.19}$$

Thus the electric field intensity vector at a point in space is defined as the force that would be exerted on a unit "test" charge placed at that point; it has the units newtons per coulomb.

In free space, or in a material that we will characterize in a following discussion as linear, the electric field due to multiple charge sources is additive. That is, if we have two point charges, the $\bar{E}$ fields of the two point charges add vectorially; i.e.,

$$\bar{E}_{\text{tot}} = \bar{E}_1 + \bar{E}_2, \tag{1.20}$$

and this of course may be extended to any number of point charges in a *charge cluster*. Consider, for example, a "cloud" of charged particles, where the individual particles are electrons or positive ions. The individual charges in this case are so small and so numerous that it becomes convenient, as far as macroscopic effects are concerned, to define a *charge density* $\rho$ such that the total charge in any small incremental volume $\Delta v$ is $\rho \, \Delta v$. That is, in the limit we may write

$$dq = \rho \, dv, \tag{1.21}$$

where $dq$ is the differential charge in the differential volume $dv$. This charge density is a point function; i.e., it may be a constant or it may be different at each point throughout the volume, in which case we would write $\rho \, (x, y, z)$.

Consider now a volume distribution of charge, with a given charge density of $\rho$ coulombs/meter³. If this volume is divided into differential volume increments $dv$, then each differential volume will contain a charge $dq = \rho \, dv$, which may be thought of as an infinitesimal point charge, and the differential electric fields from each of these differential charges add vectorially, just as in the case of a charge cluster, to give the total $\bar{E}$ field. (The addition, of course, becomes in the limit an integration.) In Figure 1.13 we have denoted the *source coordinates* by $x'$, $y'$, and $z'$ and the coordinates of the point at

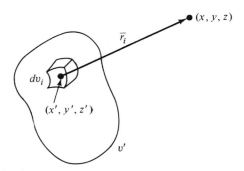

**Figure 1.13**   Field Due to a Volume Charge Distribution

which the field is being evaluated by $x$, $y$, and $z$. From the differential volume $dv_i$ we have the contribution

$$d\bar{E}(x, y, z) = \frac{\rho(x', y', z') \, dv_i' \, \bar{r}_i}{4\pi\epsilon r_i^3},$$  (1.22)

where, of course, $\bar{r}$ depends on both the primed and unprimed coordinates. To obtain the total $\bar{E}$ field, we must integrate over all the source volume (primed coordinates):

$$\bar{E}(x, y, z) = \int_{x'} \int_{y'} \int_{z'} \frac{\rho(x', y', z') \, dv' \bar{r}}{4\pi\epsilon r^3}.$$  (1.23)

(In some instances, the charge distribution will be very close to the surface of the volume. For example, if a charge is placed on a good conductor such as copper, it may be shown theoretically and experimentally that the charge in equilibrium will be distributed very close to the surface of the copper. In this case it is convenient to define a surface charge density, and the integral of equation 1.23 becomes an integral over a surface rather than over a volume. The general theory, however, is the same.)

### Example 1.4

Suppose that charge is distributed uniformly over a large plane area, with a surface density $\rho_a$ coulombs/meter$^2$. We wish to compute the electrical field intensity at a distance $y$ from the plane in a dielectric material of permittivity $\epsilon$. We shall assume that the dimensions of the plane are large compared to the distance $y$, so that negligible error is made by assuming the plane to be infinite in extent. Consider a circular ring in the plane as indicated in Figure 1.14. The charge in each little increment around the ring is $dq = \rho_a x \, d\phi \, dx$. If we consider the increments in diameterically opposite pairs, it is clear that the components of the field parallel to the plane all cancel, so the field will have only a $y$ component (normal to the plane). For each increment the normal contribution to the field is

$$dE_y = \frac{dq(\cos\theta)}{4\pi\epsilon r^2} = \frac{\rho_a x \, d\phi \, dx \cos\theta}{4\pi\epsilon r^2}.$$  (1.24)

Thus the total field will be

$$E_y = \int_0^\infty \int_0^{2\pi} \frac{\rho_a x \cos\theta}{4\pi\epsilon r^2} \, d\phi \, dx$$

$$= \frac{\rho_a}{2\epsilon} \int_0^\infty \frac{x \, dx \cos\theta}{r^2}.$$  (1.25)

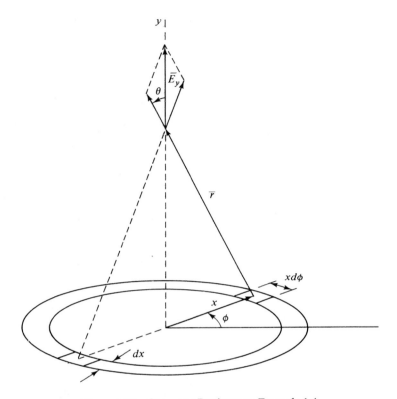

**Figure 1.14** Geometry Pertinent to Example 1.4

Since $x = y \tan \theta$, $dx = y \, d\theta/\cos^2 \theta$ and $r^2 = y^2/\cos^2 \theta$. Hence

$$E_y = \frac{\rho_a}{2\epsilon} \int_0^{\pi/2} \frac{x}{y} \cos \theta \, d\theta$$

$$= \frac{\rho_a}{2\epsilon} \int_0^{\pi/2} \sin \theta \, d\theta = \frac{\rho_a}{2\epsilon}. \tag{1.26}$$

## *Example 1.5*

Suppose that another plane carrying a uniform charge density of $-\rho_a$ coulombs/meter$^2$ is placed parallel to the plane of Example 1.4. The fields due to the charge distributions on the two planes add; between the planes the two fields have the same sign, while everywhere else the two fields have opposite signs. Hence between the planes the field intensity is

$$E = \frac{\rho_a}{\epsilon} \qquad \text{(normal to the planes)}$$

and is zero elsewhere. (Of course, in reality the planes can never be infinite in extent, but if their area is large compared with the distance between them, the above expression is a good approximation for the field intensity at points not too close to the edge of the configuration.)

### Gauss's Law

Imagine a spherical surface of radius $R$ with a point charge of $q$ coulombs at its center, as in Figure 1.15. Since the electric field is everywhere normal to the surface, and has spherical symmetry,

$$\int_s \bar{E} \cdot d\bar{s} = 4\pi R^2 E = 4\pi R^3 \frac{q}{4\pi\epsilon R^2} = \frac{q}{\epsilon}. \qquad (1.27)$$

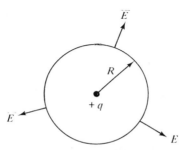

**Figure 1.15** Point Charge and Concentric Spherical Surface

We note that the flux of $\bar{E}$ through the surface is independent of $R$. This is because the area subtended by any solid angle increases as $R^2$, while the field intensity decreases as $1/R^2$. Therefore, we may distort the sphere with a spherical segment having a different radius, as indicated in Figure 1.16, without changing the result. In fact, we may allow any number of such distortions, as indicated in Figure 1.17(a), and the surface integral will remain the same. In Figure 1.17(b) we have let our little spherical segment distortions approach a limit, and in the limit it is conceptually evident that the

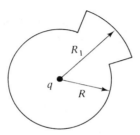

**Figure 1.16** Sphere Distorted by Spherical Segment

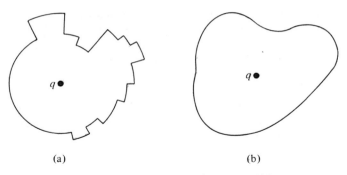

(a)                                    (b)

**Figure 1.17**  Further Distortions of the Sphere

surface integral for any shape surface will equal $q/\epsilon$, regardless of the position of $q$, provided that it is inside the closed surface. Since, in a linear medium, the fields due to multiple charges add, we may have $n$ point charges inside the surface, and the surface integral will be

$$\int_s \overline{E} \cdot \overline{ds} = \frac{1}{\epsilon} \sum_{i=1}^{n} q_i = \frac{q_{tot}}{\epsilon}. \tag{1.28}$$

Finally, it is not necessary that the charges be point charges. If the surface is filled with a volume density charge of $\rho$ Coulombs/meter$^3$, we may conceive of this volume charge as composed of many small volume increments, each having a charge $dq = \rho \, dv$, as we did previously (equations 1.21 and 1.22) in computing the electric field. Since each increment essentially acts like a differential point charge, we have

$$\int_s \epsilon \overline{E} \cdot \overline{ds} = q_{tot} = \int_v \rho \, dv. \tag{1.29}$$

(Here we have multiplied both sides of our equation 1.27 by $\epsilon$ and put the $\epsilon$ inside the integral. The reason for this is that we may wish to apply this result in configurations where $\epsilon$ is not the same at all points on the surface.) Equation 1.29 is one form of *Gauss's law*, and the use of this law often greatly simplifies the work involved in finding the electric field in problems involving symmetry, as compared to a direct integration using Coulomb's Law. This is illustrated in several of the problems at the end of the chapter.

### Voltage

We shall consider next the work required to move a charged particle in an electric field. If the particle carries a positive charge, $+q$, then it has a force $\overline{F} = q\overline{E}$ exerted on it by the field. If an external agent causes the particle to move antiparallel to the direction of the field (i.e., in a direction opposite to that of the field), the work required is obtained by simply integrating

*F dl* over the distance moved. If the particle is caused to move perpendicular to the field direction, no work is done, since the motion is then at right angles to the force vector. In general, the work is given by

$$W = -\int_a^b \overline{F} \cdot \overline{dl} = -q \int_a^b \overline{E} \cdot \overline{dl}, \qquad (1.30)$$

and this represents work done on the charge by the external agent, so that if $W$ is positive, the potential energy of the charge-field configuration is increased. If the field is conservative (as any electrostatic field may be shown to be), the work done in moving the charge around any closed path is zero.

We are now ready to define the concept of *voltage* or *potential difference*. The voltage or potential of point $b$ with respect to point $a$ is defined as the work required to move a unit test charge from point $a$ to point $b$. Hence, from equation 1.30,

$$V_{ba} \triangleq -\int_a^b \overline{E} \cdot \overline{dl}. \qquad (1.31)$$

To reemphasize, this is the *voltage of point* b *with respect to point* a; the voltage of point $a$ with respect to point $b$ will be just the negative of this—i.e.,

$$V_{ab} = -V_{ba}. \qquad (1.32)$$

It should be noted that two points are required to define a voltage or potential difference (although sometimes one of the points may be at infinity). For this reason, voltage is sometimes referred to as an "across" variable—it is a quantity that describes something at "one place" in a field, circuit, or device with respect to some "other place." One may talk about the voltage between two coordinates of a field in free space, or the voltage between two points in a material medium, or the voltage between two points in an electrical network; in all cases, the definition is the same (equation 1.31), and in order to have a voltage exist between two points, there must be an electric field of such a nature that the integral in equation 1.31 is not zero. Of course, the existence of the $\overline{E}$ field does not mean that all possible pairs of points will have a potential difference between them; if we choose a surface everywhere perpendicular to the $\overline{E}$ field, then any point on that surface will have zero potential with respect to any other point on the surface. Such a surface is called an *equipotential surface*.

### Example 1.6

A unit point charge is placed at the origin of a system of coordinates, as indicated in Figure 1.18. We know from Coulomb's law that the $\overline{E}$ field is radial and is given by

$$\overline{E} = \frac{1}{4\pi\epsilon r^2} \overline{i}_r, \quad \text{newtons/coulomb.} \qquad (1.33)$$

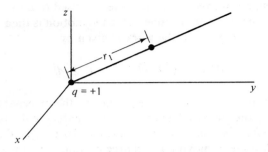

**Figure 1.18**  Potential Due to Point Charge at Origin

Let us consider a radial line from the origin to infinity and calculate the potential (voltage) of all points on this line with respect to infinity. If we consider a point at a radius $r_1$, the potential of that point with respect to infinity is

$$V = -\int_{\infty}^{r_1} \overline{E} \cdot \overline{dl} = -\int_{\infty}^{r_1} E \, dr$$

$$= -\int_{\infty}^{r_1} \frac{1}{4\pi\epsilon} \frac{dr}{r^2} = \frac{-1}{4\pi\epsilon} \left( -\frac{1}{r} \right)_{\infty}^{r_1} = \frac{1}{4\pi\epsilon r_1}. \tag{1.34}$$

Similarly, the potential of point $r_1$ with respect to another finite point, $r_2$, is

$$V_{12} = -\int_{r_2}^{r_1} E \, dr = \frac{1}{4\pi\epsilon} \left( \frac{1}{r_1} - \frac{1}{r_2} \right). \tag{1.35}$$

Since the result is the same for any radial line chosen, the equipotential surfaces will be spheres centered at the origin; as we move over the surface of any such sphere, we are moving normal to the field and hence accumulate no potential difference. In this case it is most convenient to reference all points with respect to infinity in defining their potential; this is not, however, always convenient (see Problem 1.12).

The unit of voltage or potential difference in MKS units is the volt. Since the unit of work is the joule, and voltage is the work required to transport 1 C between two points, it follows that

$$\text{volts} = \text{joules/coulomb.}$$

The unit of electrical field intensity, $\overline{E}$, thus becomes volts/meter, and this unit is equivalent to the previously used newtons/coulomb. Thus we may also write

$$\text{volts} = \text{newton-meters/coulomb.}$$

If the potential of point *a* with respect to point *b* is positive, we say that in going from *b* to *a* we experience a *voltage rise*; conversely, in going from point *a* to *b*, we see a *voltage drop*. These terms prove most convenient in later discussions of electrical networks.

### Physical Voltage Sources

There are many different types of physical devices which act essentially as voltage sources. Examples include a variety of electrostatic machines, such as Van de Graff generators; many types of electrical generators, which convert mechanical to electrical energy; batteries of many different kinds and sizes, for converting chemical to electrical energy; solar cells, which act directly upon solar energy; and so forth. It is not possible at this point to go into detail on the theory and properties of all these devices, which can be considered essentially to be devices that provide and maintain differences in potential between their terminals. (A rather old-fashioned and not very accurate term applied to the potential difference between the terminals of a source is EMF, or *electromotive force*.) It does help, however, to keep in mind that the basic function of any of the multitude of physical devices that may be called voltage sources is to establish and maintain an electric field, either by a direct separation of charges or by causing motion with respect to a magnetic field (a concept to be discussed in Chapter 2). In short, one should bear in mind that if a voltage exists between two terminals of any physical device, that voltage exists by virtue of being the negative of the line integral of an electric field along a path joining those two terminals, and (at least for static fields) it does not matter whether that path is taken through the physical device or through the space surrounding the device; the field is established by the action of external energy sources and must give a single unique value for the voltage between the terminals of the device. A mathematical idealization of the voltage source concept will be discussed in Chapter 3.

## 1.3 CAPACITIVE PHENOMENA

Consider two bodies, *a* and *b* in Figure 1.19, which are separated either in empty space or in some material medium of permittivity $\epsilon$. A charge of *q* coulombs has been transferred from *b* to *a*, leaving *a* with a net charge of $+q$ and *b* with a net charge of $-q$ coulombs. We shall assume that the surfaces of the two bodies are equipotential surfaces; i.e., all points on any one of the surfaces are at the same potential. This will be the case if these surfaces are good conductors (such as, e.g., copper), as will be discussed in Chapter 2. Since there is a separation of charge, there will be an electric field established between the two bodies, and therefore a voltage between the two bodies, given by the negative line integral of the electric field between any two points on the two surfaces, connected by some path. If the two bodies are in free

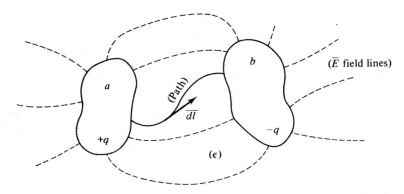

**Figure 1.19**  Two Bodies Separated in a Dielectric Medium

space or in a linear dielectric material, then at all points the $\bar{E}$ field will have a magnitude proportional to the amount of charge transferred between the two bodies, and, similarly, the voltage between the two bodies will be proportional to the charge $q$. That is,

$$q = -\int_b^a C\bar{E}\cdot\overline{dl} = CV_{ab}, \qquad (1.36)$$

where $C$ is the constant of proportionality relating the charge to the potential difference. This constant is called the *capacitance* of the configuration and has the units

$$\text{coulombs/volt} \triangleq \text{farads.}$$

It should be noted that capacitance is a property of the geometry of separated equipotential surfaces and the material comprising the separating medium. Thus a configuration may have a capacitance regardless of whether or not a separation of charges has actually taken place; one need only have a situation where the charge separation is conceivable. (For example, one may calculate the capacitance between the earth and the moon—the capacitance is a geometric parameter applicable to a configuration that will allow a charge separation conceptually. The fact that it may be difficult to *establish* the electric field between the earth and the moon is immaterial.)

### Example 1.7

Consider two parallel planes of infinite extent, as in Example 1.5. We have calculated that the electric field intensity at all points between these planes is

$$\bar{E} = \frac{\rho_a}{\epsilon}\bar{i}_n, \qquad (1.37)$$

where $\rho_a$ is the surface charge density in coulombs/meter$^2$ and $\bar{i}_n$ is a unit

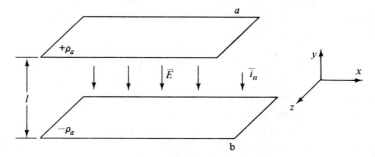

**Figure 1.20**   Parallel-plate Capacitor

vector normal to the planes in a direction as indicated in Figure 1.20. Since the field between the planes is uniform, the voltage or potential difference is

$$V_{ab} = -\int_b^a \overline{E} \cdot \overline{dy} = \frac{\rho_a l}{\epsilon},$$   (1.38)

where $l$ is the separation distance. Since, in an area $A$, there is a charge of $\rho_a A$ coulombs, an area $A$ of the configuration will have a capacitance

$$C = \frac{q}{V} = \frac{\epsilon \rho_a A}{\rho_a l} = \frac{\epsilon A}{l} \qquad \text{farads},$$   (1.39)

or the capacitance per unit area of the configuration is

$$\frac{C}{A} = \frac{\epsilon}{l} \qquad \text{farads/meter}.$$   (1.40)

### Energy Stored in a Capacitive Configuration

Whenever charges are separated, a certain amount of potential energy is stored. We will now calculate the energy stored in a capacitive configuration where a charge of $q$ coulombs has been transferred from one plate of the capacitor to the other. If the voltage between the two plates is $v$, then to transfer an incremental charge $dq$ coulombs requires an amount of work given by

$$dw = -dq \int_b^a \overline{E} \cdot \overline{dl} = v \, dq.$$   (1.41)

Since $q$ and $v$ are related by

$$q = Cv,$$   (1.42)

$dq$ may be written as

$$dq = C \, dv.$$   (1.43)

Thus the work required to bring up the voltage on the configuration from zero to a voltage of $V$ volts is

$$w = C \int_0^V v \, dv = \tfrac{1}{2} C V^2. \tag{1.44}$$

Thus one might define a *capacitor* (which is the configuration itself which displays the parameter *capacitance*) as a physical configuration, the primary function of which is to store energy in an electric field. The most common type of capacitor is constructed from thin sheets of metal foil, separated by thin dielectric sheets of specially treated paper, or, in higher-quality capacitors, mica. Air may also be the dielectric, as in variable tuning capacitors used in radio. These capacitors are all variations on the parallel-plane idea, so the formula given by equation 1.39 is important. Of course, no physical capacitor can have an infinite area; there will always be a fringing effect at the edges of the physical capacitor, so equation 1.39 is always an approximation—it is a good approximation, however, if the area $A$ is large compared with the separation $l$.

A capacitor is a device built essentially to store energy in the electric field. It must be emphasized, however, that whenever one has conducting materials separated by dielectrics, one has a configuration that exhibits capacitance. For example, two wires in an electrical circuit will have a capacitance between them, and this capacitance may be unwanted and detrimental to the operation of the circuit. Such an unwanted capacitance, the effects of which are often difficult to evaluate, is called a *stray* or *parasitic capacitance*, and we will discuss parasitics later in connection with lumped networks. In short, a capacitor is a device specifically built to have the parameter capacitance; this parameter, however, is always present in an electrical circuit or device, and is an important parameter whether one is talking about a large transmission line or a microelectronic circuit.

### Dielectric Materials and Polarization

Let us consider again a capacitor made up of two parallel plates in free space, having a permittivity $\epsilon_0 = 8.854 \times 10^{-12}$ farad/meter (F/m). As we have already seen, this configuration has a capacitance per unit area of

$$\frac{C}{A} = \frac{\epsilon_0}{l} = \frac{8.85}{l} \times 10^{-12} \ \text{F/m}^2. \tag{1.45}$$

It may be determined experimentally that this capacitance can be greatly increased by filling the space between the plates with certain types of materials, called *dielectrics*, which have a permittivity greater than $\epsilon_0$. Such materials are characterized by certain properties. First, the negative and positive charges in each atom are tightly bound within the framework of each atom—

there is no free charge available to move about from atom to atom and molecule to molecule (as in the case in most metals, which we shall discuss later). Second, the centers of positive and negative charge in each atom may be separated within the atomic or molecular framework as indicated in Figure 1.21. Such a separated charge configuration is called a *dipole*. If a dipole is placed in an electric field such that it makes an angle $\theta$ to the direction of the field (Figure 1.22), then the dipole will experience a *torque* given by

$$\tau = F_1 d \sin \theta = q \, d \, E \sin \theta, \qquad (1.46)$$

and the dipole itself is said to possess a *dipole moment*,

$$p \triangleq qd. \qquad (1.47)$$

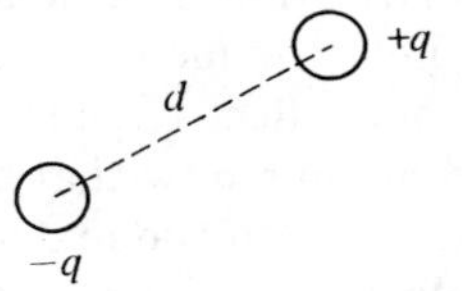

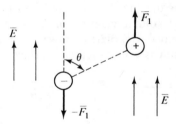

**Figure 1.21** Electric Dipole

**Figure 1.22** Dipole in an Electric Field

In some materials atoms are bonded into a molecular configuration in such a way that the molecules display permanent dipole moments. In the absence of an applied electric field, these dipoles have random orientation under the influence of thermal agitation, and there is no macroscopic effect detectable. If, however, the material is subjected to an externally applied electric field, then a torque is present on each dipole which tends to overcome the thermal agitation and align the dipoles with the field. The degree to which the thermal agitation is overcome depends on the applied field strength in a nearly linear way in most materials. Such a material, characterized by permanent molecular dipole moments, is called a *polar material*. It should be noted that in order for the dipole alignment to take place, the material must usually be a liquid or a gas; in most solids, the bonding in a crystalline structure is too strong to permit realignment of the molecules. (Certain polar liquids may be solidified while subjected to an electric field, causing the "freezing" of the dipoles in an aligned configuration; such a polarized solid is called an *electret*.)

Most dielectric materials are nonpolar; rather than having permanent molecular dipole moments, the dipoles are formed within each atom or molecule by the forces on the atomic charges resulting from the applied electric field, with the amount of charge separation approximately linearly related to the field strength. Thus, for each atom,

$$p \sim E,$$

where $E$ is the strength of the electric field in the vicinity of that atom. The macroscopic effect of the polarization of a polar and nonpolar dielectric is the same, but the nonpolar dielectric may be a solid.

To consider the macroscopic effect of polarization, let us return to our parallel-plate capacitor, and assume now that the space between the plates is filled with a polarizable material. If the material is homogeneous and *isotropic* (i.e., has the same polarization properties in all directions), then the uniform electric field will cause all atoms of the material to polarize equally, both in magnitude and direction, as indicated in Figure 1.23. At all points internal to the dielectric the "heads" and "tails" of the adjacent dipoles cancel, so internally no net volume charge density is apparent. At the surfaces of the dielectric (adjacent to the plates), however, this cancellation does not take place, and the result is a net apparent surface charge on the dielectric

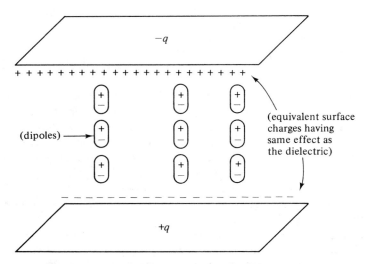

**Figure 1.23** Dielectric Between Parallel Charged Plates

surfaces. In fact, the total effect of the dielectric is exactly the same as if the material itself were removed and replaced by the two surface charge layers— as indicated in Figure 1.23. These surface charge layers have the effect of decreasing the electric field within the dielectric, since they have signs opposite to that of the charge on adjacent plates. Hence, for a given amount of free charge on the plates, the bound surface charge on the dielectric acts to reduce the field intensity, and hence the voltage between the plates, thereby increasing the capacitance. Since the electric field in the material is decreased, for the same amount of free charge, the permittivity of the material is greater than that of free space. The ratio of the permittivity of the dielectric material to the permittivity of free space is called the *relative permittivity* or *dielectric constant k*; i.e.,

$$\epsilon = K\epsilon_0. \tag{1.48}$$

The dielectric constants of some typical materials are indicated in Table 1.1.

**Table 1.1**  Typical Dielectric Constants

| Medium | K |
|---|---|
| Free space | 1 |
| Petroleum oil | 2.2 |
| Quartz | 3.8 |
| Mica | 5.4 |
| Pure water | 81 |
| Barium titanate | About 10,000 |

## PROBLEMS

**1.1**  For the vectors given below, find $\bar{A} \cdot \bar{B}$ and $\bar{A} \times \bar{B}$.
   (a) $\bar{A} = 2\bar{i}_x$,  $\bar{B} = 5\bar{i}_x$.
   (b) $\bar{A} = 2\bar{i}_x$,  $\bar{B} = 5\bar{i}_y$.
   (c) $\bar{A} = -2\bar{i}_x$,  $\bar{B} = 3\bar{i}_y - 4\bar{i}_z$.
   (d) $\bar{A} = \bar{i}_x + \bar{i}_y + \bar{i}_z$,  $\bar{B} = 2\bar{i}_x - \bar{i}_y - 3\bar{i}_z$.

**1.2**  Show that $(\bar{A} \times \bar{B}) \cdot \bar{C}$ represents the volume of a parallelepiped, where $\bar{A}$, $\bar{B}$, and $\bar{C}$ are concurrent sides.

**1.3**  Show that $\bar{A} \times (\bar{B} \times \bar{C}) = (\bar{A} \cdot \bar{C})\bar{B} - (\bar{A} \cdot \bar{B})\bar{C}$ and that

$$\bar{A} \times (\bar{B} \times \bar{C}) \neq (\bar{A} \times \bar{B}) \times \bar{C}.$$

**1.4**  If $\bar{A}$ is a vector, interpret geometrically the meaning of $d\bar{A}/dt$. Show that

$$\frac{d}{dt}(\bar{A} \cdot \bar{B}) = \bar{A} \cdot \frac{d\bar{B}}{dt} + \frac{d\bar{A}}{dt} \cdot \bar{B}.$$

**1.5** A vector field is given in rectangular coordinates by

$$\bar{V} = y\bar{i}_x.$$

Calculate the line integral of this field along the straight-line segments connecting the points $(0, 0, 0)$ and $(1, 0, 0)$, $(1, 0, 0)$ and $(1, 1, 0)$, and $(0, 0, 0)$ and $(1, 1, 0)$. What is the integral around the closed contour defined by the three segments? Is this field conservative? Can you find a contour around which the line integral is zero?

**1.6** The velocity of water in a stream is approximated by

$$\bar{V} = V_0 y\bar{i}_x,$$

where the stream bottom is specified by $y = 0$, the top by $y = 1$ m, and $\bar{i}_x$ is in the direction of flow. The stream is 10 m wide. Compute the flux (volume flow rate) through a plane surface normal to the stream bottom but making an angle of $\theta$ with the stream sides.

**1.7** Find the flux through a unit sphere centered at the origin in a uniform field.

**1.8** Two point charges, each of 1 C, are separated by a distance of 1 m in free space. Calculate the force required to maintain this separation.

**1.9** Find the energy stored in the configuration of Problem 1.8, by calculating the work required to bring one of the charges from infinity to within a distance of 1 m from the second charge (placed at the origin).

**1.10** A charge of $+q$ coulombs and one of $-q$ coulombs are separated in free space by a distance of 1 m. Show that the magnitude of the electric field intensity at any point on the plane which perpendicularly bisects a line connecting the charges is

$$E = \frac{q}{4\pi\epsilon_0 r^3},$$

where $r$ is the distance between the point and each charge. In what direction is the electric field at all points on the plane?

**1.11** In an electron beam, charge is distributed along a straight line with a linear density of $\rho$ coulombs/meter. Find the electric field intensity at a distance $r$ from the beam.

**1.12** Find the potential difference between two points, at distances of 1 m and 2 m, respectively, from the line charge of Problem 1.11. Can you find the potential of all points in the space surrounding the line charge with respect to infinity? Contrast this problem with the similar one for a point charge.

**1.13** An electron has a mass of about $9.11 \times 10^{-31}$ kilogram (kg). In a cathode-ray tube, electrons are accelerated by an electric field. If an electron starts from rest and accelerates through a potential difference of 10 kilovolts (kv), what is its terminal velocity? Its terminal energy?

**1.14** Figure 1.24 shows an electron gun and a pair of deflection plates in a cathode-ray tube. If an electron leaves the gun with a velocity $V_0$, and a voltage $v$ is

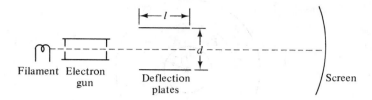

**Figure 1.24** Cathode-ray Tube

maintained between the deflection plates, describe the trajectory of the electron. What will be the deflection of a beam of such electrons from the center of the screen? (Neglect "fringing" at the edges of the deflector plates.) What sort of time variation would $v$ have to have to sweep the beam at a constant rate across the screen, as in a television picture tube?

**1.15** A spherical surface carries a uniform surface density of charge of $\rho$ coulombs/meter². Using Gauss's law, find the electric field intensity at all points, both interior and exterior to the sphere. What is the potential of all points with respect to infinity? Set up the integral that would be required to evaluate the electric field for this problem using Coulomb's law. Which of the two methods is the simpler?

**1.16** Find the capacitance of an isolated sphere of radius $R$ in free space, i.e., the capacitance between the sphere of radius $R$ and a concentric one of infinite radius.

**1.17** The sphere of Problem 1.16 is painted with a coat of dielectric paint having a thickness $l$ and a dielectric constant $K$. Find the percentage change in capacitance over that of the unpainted sphere.

**1.18** Using Gauss's law, show that the electric field intensity must be continuous at the surface of a volume charge distribution of uniform density $\rho$ coulombs/meter³.

**1.19** A sphere of radius $R$ contains a total charge of $q$ coulombs uniformly distributed throughout its volume. The space exterior to the sphere is empty. Find the electric field and potential with respect to infinity of all points, interior and exterior to the sphere. Compare with the solutions of Problem 1.15.

**1.20** A coaxial cable consists of two concentric cylindrical conductors as shown in Figure 1.25. The space between the conductors is filled with two dielectric layers as indicated. Calculate the capacitance per unit length of the cable. (*Hint:* Use Gauss's law.)

**1.21** A capacitor consists of two parallel plates of area $A$ and separation $d$ in free space. The charge on the plates is $+q$ and $-q$ coulombs, respectively. What is the energy (in joules) of the configuration? By considering this energy as a function of the separation $d$, find the force required to maintain the separation of the plates.

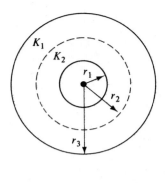

$$r_1 = 0.5 \text{ cm} \quad K_1 = 1.5$$
$$r_2 = 1.0 \text{ cm} \quad K_2 = 10.0$$
$$r_3 = 1.5 \text{ cm}$$

**Figure 1.25**  Coaxial Cable

**1.22**  A variable tuning capacitor, which is often used in radio circuits, is indicated in Figure 1.26. Neglecting all fringing effects, determine how the capacitance varies with the displacement angle $\theta$. Assuming a charge of $q$ coulombs on the capacitor, find the voltage and the energy stored as a function of $\theta$. Also calculate the torque which tends to rotate the movable plates, as a function of $\theta$.

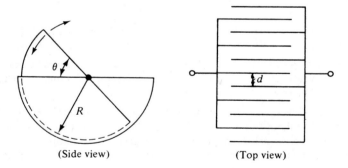

(Side view)          (Top view)

**Figure 1.26**  Variable Tuning Capacitor

# 2

# Current and
# Magnetic Field Phenomena

In Chapter 1 we concentrated essentially on theory derived from Coulomb's law, which describes forces existing between separated, stationary charges. Of course, over the time interval during which a charge separation is being established (as in the charging of a capacitor, for example), one has charges in relative motion with respect to the observer. In such cases, Coulomb's law, by itself, does not give the correct result for observed forces, and one postulates a magnetic field to account for the discrepancies. In this chapter we turn our attention to charges in motion and concepts associated with magnetic fields.

## 2.1  ELECTRICAL CURRENT

### Moving Charge and Current

Let us consider a uniform beam of moving charge, as indicated in Figure 2.1.

The beam has a cross-sectional area of $A$ meters and a uniform volume density of charge, $\rho$ coulombs/meter$^3$. The beam is moving with a constant velocity $\bar{u}$ meters/second with respect to the observer. During a time inter-

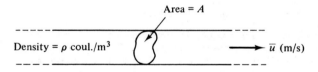

Area = $A$

Density = $\rho$ coul./m$^3$      $\bar{u}$ (m/s)

**Figure 2.1**  Uniform Beam of Moving Charge

val of $T$ seconds, a total charge given by

$$q = \rho A u T \quad \text{coulombs} \tag{2.1}$$

passes through any cross-sectional area of the beam. Hence the observer sees a flow rate of

$$I \triangleq \frac{q}{T} = \rho A u \quad \text{coulombs/second,} \tag{2.2}$$

and this rate at which charge flows through a cross section of the beam is called *current* and has the units amperes (coulombs/second). [Of course, one might consider the charge to be "stationary" and the observer in motion with a velocity $(-\bar{u})$; the observer would "see" the same current, because it is only the relative motion of the charge with respect to the observer that matters.]

Now let us suppose that the flow rate of charge in the beam is not constant, but varies with time. One may still define the current, but now its value must be defined at every instant of time; i.e., the current is now a function of time. At time $t$ we shall denote the total charge that has passed through a cross section of the beam by $q(t)$. Then, at time $t + \Delta t$, the total charge that will have passed will be $q(t) + \Delta q(t)$, and we may define the instantaneous current as

$$i(t) \triangleq \lim_{\Delta t \to 0} \frac{\Delta q(t)}{\Delta t} = \frac{dq(t)}{dt}. \tag{2.3}$$

Conversely,

$$q(t) = \int_{-\infty}^{t} i(\tau) \, d\tau; \tag{2.4}$$

i.e., the total charge that has passed through the cross-sectional area at time $t$ is the integral of $i(t)$ over all past time.[1]

---

[1]A word is in order here about the use of definite integrals in this text rather than the use of indefinite integral notation. We might also have written equation 2.4 as

$$q(t) = \int i(\tau) \, d\tau + \text{(constant)},$$

but to achieve any numerical result, the constant of integration must be evaluated, and the definite integral of equation 2.4 seems conceptually easier to relate to the physical situation. Of course, it may be argued that no physical configuration could have been in existence for $t \longrightarrow -\infty$, and hence equation 2.4 has no real physical meaning. Perhaps the best expression is

$$q(t) = \int_{t_1}^{t} i(\tau) \, d\tau + q(t_1),$$

where $q(t_1)$ is the *initial condition* or *initial value* of charge at some time $t_1$. In this text we will use this form and also the form of equation 2.4; when the lower limit is $-\infty$, it simply indicates that the entire past history of the integrand is to be included in the integration, so

$$q(t) = \int_{t_1}^{t} i(\tau) \, d\tau + q(t_1) = \int_{t_1}^{t} i(\tau) \, d\tau + \int_{-\infty}^{t_1} i(\tau) \, d\tau = \int_{-\infty}^{t} i(\tau) \, d\tau.$$

Note that the current $I$ or $i(t)$ is a *through variable,* in contrast to voltage, which is often termed an *across variable,* i.e., something defined in terms of two separated points.

### Current Density

Instead of the cross-sectional area indicated in Figure 2.1, let us now consider a plane area whose normal is at an angle of $\theta$ with respect to the velocity vector, as in Figure 2.2. Since there is no buildup of charge anywhere

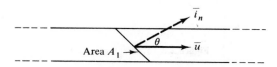

**Figure 2.2** Another Uniform Charge Beam

along the beam, the same amount of charge per unit time must be transported through this area as through the cross-sectional area previously considered. The area $A_1$, however, is now greater than $A$ by the factor $1/\cos\theta$, so the expression for the flow rate is

$$I = \rho u A_1 \cos\theta. \tag{2.5}$$

We would now like to generalize this result to apply to any surface (not necessarily a plane surface) cut by the beam. This may be done by defining a *current density vector,* $\bar{j}$, such that the total current $dI$ flowing through any incremental area $\overline{ds}$ is given by

$$dI = \bar{j}\cdot\overline{ds} = j\,ds\cos\theta \qquad \text{amperes.} \tag{2.6}$$

Hence, through any finite area, the total current is

$$I = \int_s \bar{j}\cdot\overline{ds} \qquad \text{amperes,} \tag{2.7}$$

or

$$i(t) = \int_s \bar{j}(t)\cdot\overline{ds}, \tag{2.8}$$

for a time-varying current density. The units of the current density vector are amperes/meter$^2$.

---

[2]Usually, throughout this text, constant quantities will be denoted by capitals and time-varying quantities by lower case letters.

### Current Elements

Consider now a thin filament of current, of cross section $A$, as indicated in Figure 2.3. The cross-sectional area of this filament will be assumed small enough that the current density is essentially constant over this area. The

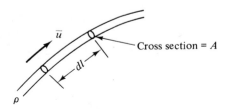

**Figure 2.3** Current Filament

current, as in equation 2.2, is

$$I = \rho A u \qquad \text{amperes,} \qquad (2.9)$$

and the total charge in the cylinder of length $dl$ is

$$dq = \rho A \, dl \qquad \text{coulombs.} \qquad (2.10)$$

Multiplying $dq$ by $u$ yields

$$dq \, u = \rho u A \, dl = I \, dl, \qquad (2.11)$$

or, in vector form,

$$dq \, \bar{u} = I \, \overline{dl}. \qquad (2.12)$$

Equation 2.12 is interesting, because it relates a moving charge to current on a microscopic or point basis. In other words, an incremental charge moving at a certain velocity may be equated to a current times an incremental distance, measured in the same direction as the velocity vector. The quantity indicated by either side of equation 1.22 is called a *current element*, or, more exactly, a *differential current element*. The current element has the dimensions coulomb-meters/second or ampere-meters, and equation 2.12 has important implications, as we shall see, in defining the magnetic field.

Suppose now that instead of an incremental amount of charge, $dq$, we have a finite charge $q$ concentrated essentially at a point (our old friend the point charge), moving at a velocity $\bar{u}$. The product $q\bar{u}$ still has the dimensions ampere-meters, but now it is a finite quantity rather than an infinitesimal one, and we may call it a *finite current element*, from the close analogy with equation 2.12. (Of course, in this case, $I$ in equation 2.12 must be infinite if $I \, \overline{dl}$ is to be finite; this is not too strange if one considers the point charge passing through a surface. Since a finite charge is transported through a surface in an infinitesimal time, the "current" through the surface is infinite, at one instant of time, and zero at all other times.) We will return to the consideration of both differential and finite current elements when we discuss the magnetic field, in Section 2.3.

## 2.2 CONDUCTION IN MATERIAL MEDIA

An electrical current consists of charges in motion with respect to the observer. In order to have a current, then, one must have charge available for conduction. We may have an electrical current in a vacuum, or air, as in the case of our electron beam previously discussed, but by far the most usual and important currents are currents that flow in material media and, in particular, solids. In this section we shall discuss, qualitatively, how charge is made available for conduction in materials that may be classified as either good conductors or semiconductors. Good conductors, of course, have always been of primary importance in electrical engineering, while during the past twenty years semiconductors have literally revolutionized modern life, particularly in areas related to computers and information processing. This discussion must necessarily be qualitative and somewhat brief; a thorough, quantitative treatment of conductors and semiconductors must be based upon the theory of quantum mechanics and is beyond the scope (and intent) of this text.

### Atomic Assemblies and Bonding Mechanisms

The classical model of the atom, as derived from experimental evidence by Rutherford, Bohr, and others in the early nineteen-hundreds, is comprised of a nucleus containing protons and neutrons, with electrons in orbit about this nucleus, the orbital distances being great in comparison to the dimensions of the nucleus. The hydrogen atom, for example, contains one proton in the nucleus and one orbital electron. Associated with this orbital electron is a certain amount of energy, made up partly of the kinetic energy of the moving electron and partly of the potential energy due to the separation of this electron from the proton in the nucleus with its associated electrostatic field. According to classical mechanics, the orbit need not be circular in order for the total energy of the electron to be constant; an infinite number of elliptical orbits are also allowed, in which the kinetic and potential energies are continually changing with time, but in such a way that the total energy remains constant.

While this simple orbital model of the atom is sufficient to explain some experimental results, such as scattering experiments performed by Rutherford around 1911, it is not sufficient to explain others. For example, if a large potential difference is established between two electrodes in hydrogen gas, an electrical discharge will occur accompanied by the release of energy from the hydrogen in the form of light and other radiation, and this radiation occurs only at certain discrete frequencies (spectral lines). According to modern radiation theory, each frequency $f_i$ is associated with photons or *wave packets*, each having an energy

$$W_i = hf_i, \tag{2.13}$$

where $h$ is Planck's constant. Since the energy of a radiated photon corresponds to the difference in the energies of an electron prior to and after the emission of the photon, it is apparent that each electron can exist only in certain discrete energy states; i.e., the energy levels of the electron in the atomic orbit are quantized.

Even if discrete energy levels are postulated for each electron, there are an infinite number of different elliptical orbits which would yield each energy. However, the total energy of the electron is not the only thing that is found experimentally to be quantized. In fact, modern experimental evidence shows the classical mechanical model of the orbiting electron to be quite inadequate. If only the energy of the electron were quantized, then a single number (quantum number) would be required to specify the electron state; in hydrogen, for example, the energy states are given by

$$W_i = \frac{2.18 \times 10^{-18}}{n_i^2} \quad \text{joules,} \tag{2.14}$$

where $n_i$ is any integer. Experimental evidence, in fact, shows that four quantum numbers are required to specify completely the state of the electron; i.e., the electron configuration in the atom has four properties, all discrete. The first is designated by a quantum number denoted by $n$ (the principal quantum number), and this denotes the energy level, as previously described. The second property is denoted by the quantum number $l$ and (in terms of the orbital model) is related to the shape of the elliptical orbit. That is, for a given energy, only certain elliptical shapes are allowed (from circular to very elongated). For each $l$ it is found that only $(2l + 1)$ possible orientations of the ellipse in space are allowed; this is indexed by a third quantum number, $m_l$. Finally, an electron with a specified $n$, $l$, and $m_l$ is still found to have two possible states $(m_s)$, and these are thought to be associated with the spin of the electron on its axis, this spin having two possible orientations (i.e., spin axis up or down).

When atoms combine to form a molecular or crystalline structure, one must consider the various energy levels of the system of atoms as a whole, and quantitatively this problem is very complex, as is evident when one considers that each electron in the system requires four quantum numbers to describe its state. However, we can easily answer the question of why atoms bond together to form molecules or crystals in the first place; atoms bond together in a more-or-less stable configuration when the *energy of the system is less than the energy associated with a separated configuration.* Nature, in other words, tries to form configurations that have a minimum of stored energy, and this may occur with systems composed of either similar or dissimilar atoms. Atoms which when in proximity of each other form a configuration having less stored energy than when they are separated are said to form bonds, and the bonding mechanisms of primary importance are given special names: *metallic bonding, covalent bonding,* and *ionic bonding.*

### Conduction in Metals: Resistance

Most metals are excellent conductors of electrical current, and conversely, most good conductors are metals. Most metals are characterized by having a small number of electrons (usually one or two) in the outermost orbit—the *valence electrons.* In copper, for example, all the energy states for $n = 1, 2,$ and 3 (i.e., the three inner rings or orbits) are filled, and there is a single electron in the fourth, or valence ring, corresponding to an energy state $n = 4$. The inner core of the atom, comprising the nucleus and the three inner electron energy levels, is known to be about $1.27 \times 10^{-10}$ m in radius, and when copper atoms combine to form a crystalline structure, the nearest nuclei are known to be separated by about $2.55 \times 10^{-10}$ m. In other words, if we consider the outermost electron, in a highly elliptical orbit around the nucleus, it is possible for this electron on the average to be closer to a nucleus than in the case of an isolated atom. This is because it spends much of its time closer to the nucleus of an adjacent atom than to the nucleus of the parent atom. The atoms of copper thus tend to form a crystal that maximizes this effect or minimizes the net energy of the crystalline structure. This structure, for copper, consists of a cubic lattice, with atoms at each corner of a unit cube and at the center of each face of each unit cube; this same structure is typical of many other metals, such as silver, gold, and aluminum.

Because of the close atomic spacing in metallic crystals, we may visualize the material as containing a cloud or plasma of charges (electrons), which, since they are attracted nearly as much by neighboring nuclei as by the nuclei of the parent atoms, are very loosely bound to the crystalline struture. In fact, the charge density represented by this plasma is enormous; 1 cm³ of copper contains about $8.5 \times 10^{22}$ essentially free electrons, or about $-13.6 \times 10^3$ C of charge, capable of moving readily under the influence of an applied electric field.

In the absence of an applied electric field, the charge, and in fact the atoms themselves, are not at rest, but have random vibratory motions that produce the phenomena of heat. One may picture each atom as being held elastically in the crystal lattice but vibrating about its nominal position in the lattice with an amplitude nearly proportional to the temperature of the metal (thermal agitation). As a free electron approaches an atom, it may transfer some of its energy to that atom and become temporarily trapped in an orbit representing a lower-energy state for that electron. As the general configuration changes, this electron may escape again, with a random initial velocity imparted by the vibrating atom. Such a random path for an electron is indicated in Figure 2.4.

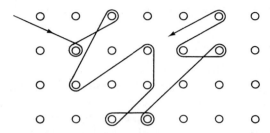

**Figure 2.4** Random Electron Path in Crystal Lattice

If we visualize a surface within the material, the charge that is transported across this surface per unit time averages to zero (even for very short time intervals)—hence the current, in a macroscopic sense, is zero. If, however, an electric field is applied to the configuration, each electron, during its period of free flight between atoms, will acquire an additional velocity due to the electric field, always in the direction of the field. The random captures still take place, but now there is an average drift velocity of charge in the direction of the field, and this average drift velocity is proportional to the electric field intensity. [Actually, the average drift velocity is small compared to the thermal velocities of the charge; in a copper conductor carrying a current density of 1000 amperes/inch² ($A/in^2$), the average drift velocity is about $1 \times 10^{-4}$ m/s, while thermal velocities at room temperature are of the order of $10^5$ m/s. The large current resulting from such a small drift velocity is due to the enormous charge density in the material.] The constant of proportionality between the current density and the electric field intensity in a linear material is called the *conductivity* of the material and is denoted by $\sigma$; i.e.,

$$\bar{j} = \sigma\bar{E} = \frac{1}{\rho}\bar{E}. \tag{2.15}$$

The term $\rho$, the reciprocal of $\sigma$, is called the *resistivity* of the material.

As the temperature of a metal is increased, its resistivity is found to increase, in an approximately linear fashion. If we denote the resistivity of the metal at two different temperatures by $\rho_1$ and $\rho_2$, at temperatures $T_1$ and $T_2$, respectively, we may express the relationship as

$$\rho_2 = \rho_1[1 + \alpha_1(T_2 - T_1)], \tag{2.16}$$

where $\alpha_1$ is called the *temperature coefficient of resistance* at temperature $T_1$. The increase of resistivity with temperature is due to the increased thermal agitation of the atoms in the lattice, which increases the probability of an electron being captured in any incremental time interval, thus decreasing the mean free path of the electrons. This, in turn, reduces the average drift velocity for a given applied field and hence the current density.

As kinetic energy is supplied to the electrons by the externally applied electric field, this same energy is transferred to the lattice atoms in the process of capturing those electrons; this shows up as an increase in the lattice vibrations, or heat. Suppose that in a certain region in the metal the current density is uniform; consider a small volume element of area $A$ and length $l$ in this region, oriented with its sides parallel to the direction of current flow. The magnitude of electric field intensity in this volume is

$$E = \rho j, \tag{2.17}$$

and the potential difference across the volume element is

$$v = El = \rho jl. \tag{2.18}$$

In a time interval $dt$, a charge

$$dq = jA\,dt \tag{2.19}$$

is transported across the element, and this represents an amount of work

$$dw = v\,dq = \rho j^2 Al\,dt \quad \text{joules.} \tag{2.20}$$

The *power* supplied by the field to this volume element is thus

$$p = \frac{dw}{dt} = \rho j^2 Al \quad \text{watts,} \tag{2.21}$$

or the *power per unit volume* supplied by the field is

$$\frac{\text{power}}{\text{unit volume}} = \rho j^2 \quad \text{watts/meter}^3, \tag{2.22}$$

and this power manifests itself as heat.

If two terminals of a voltage source are attached to a piece of conducting material, such as the metal that we have been discussing, an electric field will be established in the material and a current will flow between the terminals, and if the current density is proportional to the electric field intensity, then the total current will be proportional to the voltage; i.e.,

$$v = Ri \quad \text{or} \quad i = Gv, \tag{2.23}$$

where the constants of proportionality are called the *resistance* and *conductance* of the configuration, respectively. (Both equations 2.23 and 2.15 are referred to as *Ohm's law*.) In a conductor of uniform cross section, we may relate the resistance between two cross-sectional planes to the resistivity $\rho$ of the material as follows. Consider the conductor of uniform cross section as indicated in Figure 2.5. If the current density is uniform in the conductor, the total current

$$i = \int_A \bar{j} \cdot \overline{ds} = jA. \qquad (2.24)$$

**Figure 2.5**   Conductor of Uniform Cross Section

The voltage of plane $a$ with respect to plane $b$ is

$$i = -\int_b^a \bar{E} \cdot \overline{dl} = EL = \rho jL = \frac{\rho iL}{A}. \qquad (2.25)$$

Therefore,

$$R \triangleq \frac{v}{i} = \frac{\rho L}{A} \qquad \text{volts/ampere, or ohms.} \qquad (2.26)$$

The unit of electrical resistance is called the *ohm*, which means that the unit of resistivity is the *ohm-meter*. It should be carefully noted at this point that *resistivity* is a property of the material alone, whereas *resistance* is a property dependent not only on the material but also on the configuration (i.e., dimensions, shape, etc.).

If we have a configuration such that a voltage $v$ may be applied between a pair of terminals in the conducting medium, and if a current $i$ flows between these terminals, then in a time interval $dt$ the charge transported between the terminals is $dq = i\,dt$, and the work supplied to this charge is

$$dw = vi\,dt \qquad \text{joules.} \qquad (2.27)$$

Thus the power supplied by the voltage source (which will be converted to heat, as previously discussed) is

$$\frac{dw}{dt} = vi \qquad \text{watts.} \qquad (2.28)$$

Note that this represents *instantaneous power*; i.e., if $v$ and $i$ are functions of time, we may write

$$p(t) = v(t)i(t) \qquad \text{watts};\tag{2.29}$$

if we desire the time average power over some time interval $(T_1, T_2)$, we may average this instantaneous power over that interval; i.e.,

$$p_{\text{av}} = \frac{1}{T_2 - T_1} \int_{T_1}^{T_2} p(t)\, dt.\tag{2.30}$$

### Semiconductors

All material media conduct electricity to some extent, but the differences in conductivity (or resistivity) between different materials can be enormous, as indicated by the following typical values of resistivity:

Mica: $\qquad \rho = 2 \times 10^{15}$ ohm-meters

Germanium: $\quad \rho = 0.4$ ohm-meter

Copper: $\qquad \rho = 1.7 \times 10^{-8}$ ohm-meter.

These three materials are examples of *insulators, semiconductors,* and *good conductors* (metals), respectively, and most materials may be readily classified in this way. In this discussion we will focus our attention on the second class of materials, the semiconductors.

The two semiconductor materials most used in engineering applications are germanium and silicon. Both are crystalline in structure, with a shiny metallic appearance. Both are characterized by four electrons in the outer-most (valence) shell, in contrast to metals, which typically have only one or two valence electrons, and insulators, whose valance shell is nearly filled. Such atoms do not form the tightly packed crystal lattice typical of metals; in fact, the bonding mechanism is quite different from the metallic bonding previously discussed and is called *covalent bonding.*

Covalent bonding is a term applied to a phenomenon whereby two atoms share a pair of valence electrons, in such a way that the energy of the combination is less than that of the separate atoms. Not only semiconductors form such covalent bonds; hydrogen atoms, for example, share electrons in this way to form a stable $H_2$ molecule, as indicated by the simplified orbital model of Figure 2.6. The decreased energy of this covalent configuration is largely due to the fact that when an electron is roughly in a region between two nuclei, it is attracted by both, and the potential energy is decreased. However, it is also repelled by the other electron, and the two nuclei also repel each other, so the reason for the lowered energy is not at all clear and really requires a detailed quantum mechanical analysis. In fact, quantum

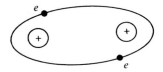

**Figure 2.6**   Orbital Model of $H_2$ Molecule

theory tells us that only pairs of electrons having opposite spins can form a covalent bond.

In germanium and silicon, a crystalline lattice is formed such that each of the four valence electrons shares an electron with an adjacent atom in a covalent bond; thus each atom may be thought of as being connected to four other atoms, as indicated diagrammatically in Figure 2.7. As long as

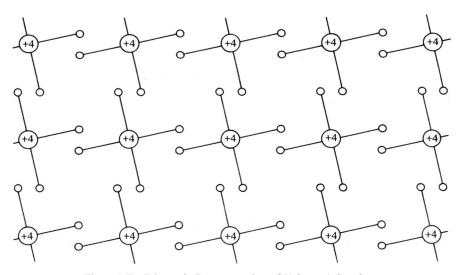

**Figure 2.7**   Schematic Representation of Valence 4 Covalent
Bonding

each electron is held in the lattice in a covalent bond, there is no charge available for conduction. Two mechanisms, however, may be responsible for making charge available for conduction. Depending upon which mechanism predominates, the semiconductor is called either an *intrinsic semiconductor* or an *impurity semiconductor*.

In an intrinsic semiconductor, thermal agitation is responsible for breaking some of the covalent bonds. If a bond is broken and an electron escapes from the lattice structure, that electron becomes free and mobile. At the same time, it leaves behind a "hole," i.e., two atoms that need an electron to re-establish the covalent bond. An electron from elsewhere in the lattice may move into this hole, annihilating the hole but leaving another hole where it came from. Thus not only are the freed electrons able to move and provide current, but one may consider the holes to move as well, in the same sense that a gap in a line of traffic moves as one car after another moves ahead to fill the gap. The apparent hole movement in fact appears as a propagation of positive charge—i.e., a hole behaves like a particle having the same mass as an electron but opposite charge. However, since the mechanism of movement of the free electron and the hole are quite different (many electrons need to move for the hole to propagate), the mobility (which may be precisely defined) of the hole and the electron are very different. The density of charge in the semiconductor is also much different from that in a metal. In a metal of valence 1, nearly one electron per atom is available for conduction, while in a semiconductor, only one free electron in $10^6$ or $10^8$ atoms may be available, which accounts for the difference in resistivities. Also, in an intrinsic semiconductor, as the temperature is increased, more free charge is made available by thermal agitation, but the mobility (or mean free path) of that charge is decreased for the same reason. Hence, over a given temperature range, the resistivity may either increase or decrease, depending on which effect predominates.

If minute amounts of an impurity of valence 3 or 5 is added to the silicon or germanium, electrons or holes are made available for conduction in another manner. Figure 2.8 illustrates schematically a valence 5 atom (such as arsenic or antimony) in the semiconductor lattice. Four of the electrons

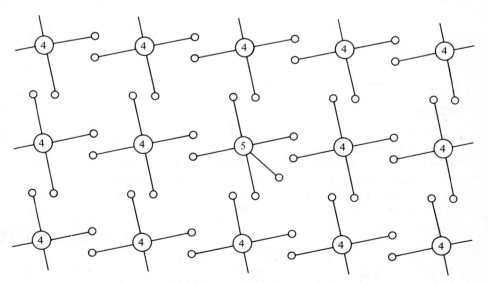

**Figure 2.8**   Valence 5 Impurity Atom in the Crystal Lattice

of this atom have formed covalent bonds with adjacent atoms, but one electron is left over and is free for conduction. A semiconductor that is doped in this way is called an *n-type semiconductor*, since the current is made up essentially of a drift of electrons contributed by the donor impurity atoms.

Figure 2.9 illustrates schematically a valence 3 atom (such as gallium or indium) in the crystal lattice. Now an electron is missing—one of the covalent bonds is not complete. Hence the effect of the impurity atom in this case is to provide a hole, and the impurity atom is called an *acceptor atom*. Since the current now consists essentially of a migration or drift of holes, which behave like positive charge, the material is called a *p-type semiconductor*.

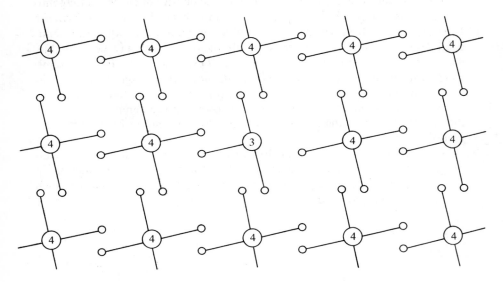

**Figure 2.9**   Valence 3 Impurity Atom in the Crystal Lattice

### The pn-Junction Rectifier

Modern electronics is almost entirely dependent upon phenomena associated with junctions between different types of semiconductor materials. Two extremely important classes of devices, rectifiers (diodes) and transistors, result from utilizing these phenomena. Although a thorough quantitative treatment of these devices is beyond the scope and intent of this text, the emphasis of which is on the circuits and systems aspects of electrical engineering, the importance of semiconductor junctions warrants a brief qualitative introduction to the subject. Hence we shall briefly discuss the semiconductor diode now and the junction transistor in Chapter 3.

Let us consider a donor- and an acceptor-type material brought together to form a junction, as indicated in Figure 2.10(a). Free electrons from the

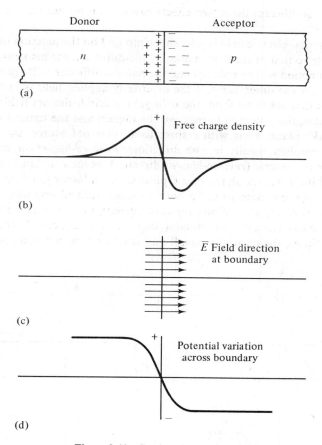

**Figure 2.10** Semiconductor Junction

donor side diffuse across the boundary, as do holes from the acceptor side. This natural diffusion of free charge across the boundary establishes free charge layers on either side of the junction [Figure 2.10(b)], which in turn establish an electric field across the boundary in such a direction as to oppose further diffusion [Figure 2.10(c)]. Thus the diffusion process reaches an equilibrium condition, with a potential distribution (negative integral of the electric field) somewhat as indicated in Figure 2.10(d). Since electrons experience a force opposite in direction to the electric field, they want to "climb up" a potential "hill", whereas holes want to "roll down" a potential "hill." Thus the potential established across the junction presents a barrier to further natural diffusion of free charge across the boundary. Of course, some high-energy electrons will still penetrate the potential barrier, and some electrons freed by thermal agitation in the *p*-type material will roll up the potential hill, but in equilibrium these two effects cancel, and no net charge is transported.

If an external electric field is now superimposed on the junction in a direction opposite to that of the field caused by the diffusion, the net field strength across the junction will be reduced and the natural diffusion will again readily take place. On the other hand, if the externally applied field is in the same direction as that resulting from the original diffusion, the net field strength will be greater, the potential barrier will be higher, and the natural diffusion does not take place. Thus with respect to an external source, the junction appears to conduct readily in one direction (*forward-biased* direction) and very little in the reverse (*reverse-biased*) direction, as indicated in Figure 2.11. Figure 2.12 indicates the shape of a typical semiconductor junction voltage–current terminal characteristic. To have an ideal rectifier, one would require zero reverse current, and infinite forward current, and of course this is just as impossible as having an ideal resistor, capacitor, or inductor. In the reverse-biased condition, the current is small and is caused by electrons and holes

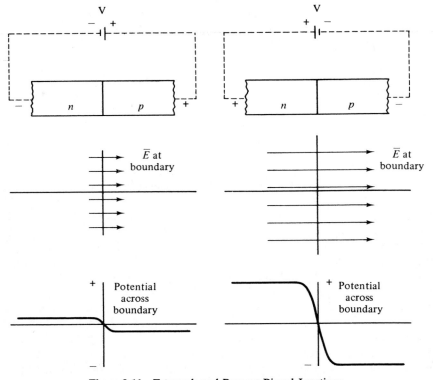

**Figure 2.11**  Forward- and Reverse-Biased Junctions

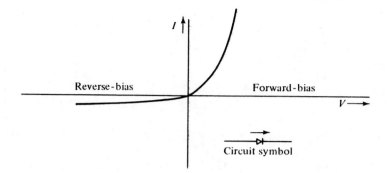

**Figure 2.12** Semiconductor Diode Characteristic

that have enough thermal energy to penetrate the potential barrier. In the forward direction, the diode appears as a somewhat nonlinear resistor of small value.

## 2.3  THE MAGNETIC FIELD

When electric charges are in motion with respect to an observer, the observed forces are no longer consistent with Coulomb's law. For example, we might consider the following three experimental facts:

1. Two point charges separated by a distance $l$ move with the same velocity $\bar{u}$ with respect to an observer. The observer "sees" a force exerted by each charge on the other, the amount of which is not consistent with Coulomb's law.
2. A point charge moves (relative to an observer) parallel to a wire carrying an electric current. An apparent force is observed between the wire and the charge. According to Coulomb's law, there should be no force, since the wire is neutral (no net charge on the wire).
3. Between two parallel wires, both of which carry current, there is an observed force; again, Coulomb's law says that there should be none, since both wires are uncharged.

### Biot–Savart Law

All these bits of experimental evidence may be explained by a law that gives a force relationship for current elements, which were discussed in Section 2.1. This law is usually called the *Biot–Savart law* and may be formulated as follows. Consider two filaments of current, as indicated in Figure 2.13. If we choose two current elements, $i_1 \, \overline{dl_1}$ and $i_2 \, \overline{dl_2}$, separated by a vector $\bar{r}$ as

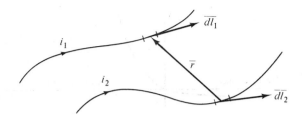

**Figure 2.13** Two Current Filaments

shown in the figure, then a differential force will be exerted on $i_1\ \overline{dl_1}$ by $i_2\ \overline{dl_2}$, and experimental evidence indicates that that force is given by

$$\overline{dF}_1 = \frac{\mu}{4\pi}\frac{i_1\ \overline{dl_1} \times (i_2\ \overline{dl_2} \times \bar{r})}{r^3}. \tag{2.31}$$

(Conversely, the force on $i_2\ \overline{dl_2}$ due to $i_1\ \overline{dl_1}$ is obtained by simply interchanging the subscripts in equation 2.31.) We note that this law, like Coulomb's law, is an inverse-square law (there is an $r$ in the numerator), but the geometry of the situation is more complicated than in the case of Coulomb's law—i.e., the direction in which the force acts is determined by the two vector products in the numerator. The constant of proportionality is also different and is written as $\mu/4\pi$ (rationalized units), where $\mu$ is termed the *permeability* of the medium and for free space is numerically given by $\mu_0 = 4\pi \times 10^{-7}$. Despite its slightly more complex structure, the law expressed by equation 2.31 may be thought of as being completely analogous to Coulomb's law, except for current elements rather than stationary point charges.

Equation 2.31 is an expression for a differential force. If we desire the (finite) force between any two finite segments of the wire, we must consider all possible pairs of current elements, and add up all the differential contributions by integrating. If we assume that the two wires form two closed circuits (so that there will be no buildup of charge anywhere), then the total force between the two wires will be obtained by integrating around both circuits, as indicated by

$$\bar{F}_1 = \frac{\mu}{4\pi} \oiint i_1 i_2\ \frac{\overline{dl_1} \times (\overline{dl_2} \times \bar{r})}{r^3}. \tag{2.32}$$

### Example 2.1

Let us investigate the *direction* of the force between two parallel wires carrying current in the same direction. We will consider two current elements as shown in Figure 2.14. The direction of $(\overline{dl_2} \times \bar{r})$ will be out of the plane determined by the two wires (the plane of the paper). The direction of $\overline{dl_1} \times (\overline{dl_2} \times \bar{r})$ will be normal to $\overline{dl_1}$, in the plane of the paper, and pointing in the direction

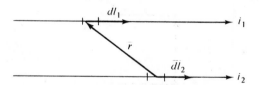

**Figure 2.14**   Two Parallel Current-carrying Conductors

of the wire carrying $i_2$. Since this will be true for any pair $\overline{dl}_1$ and $\overline{dl}_2$ that we choose, the conclusion is that there is a force of attraction between the wires. We may calculate the force of attraction per unit length for a specified separation of the wires and current values, and this will be left as a problem at the end of the chapter.

Let us consider now two point charges, moving with velocities $\bar{u}_1$ and $\bar{u}_2$ with respect to an observer. From our previous discussions it seems reasonable that the observer would "see" a total force between the charges made up of two contributions. The first contribution is the electrostatic force, as given by Coulomb's law. The second contribution to the total force is due to the fact that moving charges appear as finite current elements (Section 2.1), and hence we should expect a force given by an expression analogous to equation 2.31 to be present. The analogous expression for this force (which we will term a magnetic force, $\bar{F}_m$) is obtained by substituting $q\bar{u}$ for $i\,\overline{dl}$ in the current-element expression, to obtain

$$\bar{F}_{1m} = \frac{\mu}{4\pi}\,\frac{q_1\bar{u}_1 \times (q_2\bar{u}_2 \times \bar{r})}{r^3}, \tag{2.33}$$

where $\bar{r}$ is the distance vector as indicated in Figure 2.15. Note that this is a finite force, since it is a force between finite current elements.

**Figure 2.15**   Two Point Charges in Motion

*Example 2.2*

Suppose that two point charges move through free space with the same velocity $\bar{u}$, with respect to a stationary observer, and that $\bar{r}$ is perpendicular to $\bar{u}$, as shown in Figure 2.16. The magnetic force, as given by equation 2.33,

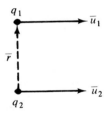

**Figure 2.16** Two Parallel Moving Charges

will be a force of attraction if both charges have the same sign, and will have a magnitude

$$F_m = \frac{\mu_0 q_1 q_2 u^2}{4\pi r^2}.$$ (2.34)

The electrostatic force between the two charges is a *repulsive* force (if both charges have the same sign), and its magnitude is given by

$$F_e = \frac{q_1 q_2}{4\pi \epsilon_0 r^2}.$$ (2.35)

The ratio of the magnetic force to the electrostatic force is

$$\frac{F_m}{F_e} = \mu_0 \epsilon_0 u^2.$$ (2.36)

It may be shown that light propagates through free space with a velocity

$$C = \frac{1}{\sqrt{\mu_0 \epsilon_0}},$$

so that the ratio of the magnetic to electric forces becomes

$$\frac{F_m}{F_e} = \frac{u^2}{c^2}!$$ (2.37)

Thus the magnetic force is small indeed compared to the electrostatic force, except for velocities approaching that of light, and in fact only assumes importance in situations (such as exemplified by our current carrying wires) when the electrostatic force is zero. (*Note:* Since $u^2 < c^2$, the net force between the two charges of equal sign is always one of repulsion.)

### Definition of the Magnetic Field Vector

To account for the force between current elements, we postulate a *magnetic field*, or *magnetic flux density vector*, $\bar{B}$, defined so that the differential force on a current element $i\,\overline{dl}$ in the field is

$$\overline{dF} = i\,\overline{dl} \times \bar{B},$$ (2.38)

or the force on a moving charge (i.e., the magnetic force) is

$$\overline{F_m} = q\bar{u} \times \bar{B}. \tag{2.39}$$

Therefore, from the Biot–Savart law, the magnetic field caused by a current element $i\,\overline{dl}$ is

$$\overline{dB} \triangleq \frac{\mu}{4\pi} \frac{i\,\overline{dl} \times \bar{r}}{r^3} \tag{2.40}$$

and that caused by a moving charge is

$$\bar{B} \triangleq \frac{\mu}{4\pi} \frac{q\bar{u} \times \bar{r}}{r^3}, \tag{2.41}$$

where $\bar{r}$ is the vector from the current element or charge to the field point at which $\bar{B}$ is being evaluated. Note that the magnetic field is always perpendicular to $\overline{dl}$, or $\bar{u}$. If we have a wire that forms a closed circuit and carries a current $i$, then to find the $\bar{B}$ field at any point in the vicinity of this wire, we must add up the contribution of each $i\,\overline{dl}$ current element around the entire circuit by integration; i.e., we may write

$$\bar{B} = \frac{\mu}{4\pi} \oint \frac{i\,\overline{dl} \times \bar{r}}{r^3}. \tag{2.42}$$

Of course, this integral may be difficult to evaluate in practice, since $\bar{r}$ changes in both magnitude and direction as a function of both the field-point coordinates and the coordinates of $\overline{dl}$. Some simple examples will be given in the problems.

### Total Force on a Moving Charge

If a charge is observed to move with a velocity $\bar{u}$ relative to the observer in a region where the observer "sees" both an $\bar{E}$ field and a $\bar{B}$ field, then equations 1.18 and 2.39 may be combined and the total force on the charge is given by

$$\bar{F} = q(\bar{E} + \bar{u} \times \bar{B}). \tag{2.43}$$

This expression is called the *Lorentz force equation* and states that the total force on the moving charge is the sum of the electrostatic force $q\bar{E}$ and the magnetic force $q\bar{u} \times \bar{B}$.

### Ampère's Circuital Law

In the case of electrostatic fields, we have seen that it is often easier to calculate the electric field using Gauss's law than it is to integrate over the charge distribution directly, using Coulomb's law. An analogous situation exists for the magnetic field; the analogous law is called *Ampère's circuital law* and applies in cases when the current is not varying with time (*stationary currents*) or at least is not varying too rapidly with time (*quasi-stationary currents*).

Consider a long, straight wire carrying a constant current $I$. The $\bar{B}$ field around this wire is everywhere tangential to a circle centered on the wire, as indicated in Figure 2.17, and has a magnitude given by

$$B = \frac{\mu I}{2\pi r}. \tag{2.44}$$

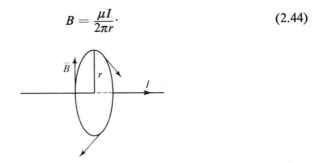

**Figure 2.17** $\bar{B}$ Field About a Straight Wire

(See Problem 2.10.) If we integrate the $\bar{B}$ vector around a circular contour of radius $r$, we obtain

$$\oint \bar{B} \cdot \overline{dl} = 2\pi r B = \mu I, \tag{2.45}$$

where $\mu$ is the permeability of the medium in which the contour is taken. Since this result is independent of the radius of the circle, we may distort this circle with as many circular segments of different radii as we choose (analogous to the way in which we distorted the sphere in discussing Gauss's law), as illustrated in Figure 2.18. Carrying this process to the limit, it becomes evident that

$$\oint \bar{B} \cdot \overline{dl} = \mu I \tag{2.45}$$

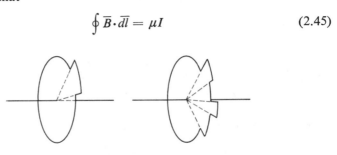

**Figure 2.18** Distorted Circular Contour

for a contour of any shape, in a plane normal to the wire. If more than one wire threads the contour, then the fields will add, so that $I$ in equation 2.45 would then represent the total current passing through the surface bounded by the contour. Finally, in the most general case, it seems logical that if we have a surface of any shape and orientation in a region where there is a current density $\bar{j}$, then

$$\oint \overline{B} \cdot \overline{dl} = \mu \int_s \bar{j} \cdot \overline{ds} = \mu I_{tot}. \tag{2.46}$$

This should not be considered a proof of Ampère's circuital law (equation 2.46), only a demonstration of its reasonableness. Also, it should be re-emphasized that if very rapidly time varying currents or charges are involved, $I_{tot}$ must be interpreted as including another type of current, called *displacement current*, which we shall not discuss here.

### Example 2.3

A long, straight conductor of radius $R$ carries a uniform current density $\bar{j}$ throughout its cross-sectional area. Find the $\overline{B}$ field both inside and outside the conductor.

Outside the conductor ($r > R$), $B$ will be the same as if all the current were concentrated in a filament; i.e., using Ampère's circuital law,

$$\oint \overline{B} \cdot \overline{dl} = 2\pi r B = \mu I = \mu(\pi R^2 j). \tag{2.47}$$

Hence

$$B = \frac{\mu R^2 j}{2r} \qquad (r > R). \tag{2.48}$$

Inside the conductor ($r \leq R$), we have

$$\oint \overline{B} \cdot \overline{dl} = 2\pi r B = \mu(\pi r^2 j).$$

Hence

$$B = \frac{\mu j r}{2} \qquad (r \leq R). \tag{2.49}$$

(The direction of $\overline{B}$ is in both cases tangential to a circle of radius $r$, in a right-hand sense.)

### Magnetic Flux

If we have a surface defined in a magnetic field $\overline{B}$, we may find the flux of this field through the surface by forming the surface integral

$$\phi = \int_s \overline{B} \cdot \overline{ds} \qquad \text{webers.} \tag{2.50}$$

The unit of magnetic flux is the *weber* (newton-meter-second/coulomb), and the unit of $\bar{B}$ is therefore *webers/meter²* (or newton-second/coulomb-meter). It is experimentally evident that the magnetic flux through any closed surface in a magnetic field is zero, in contrast to the electric field flux, which is not zero over a closed surface if that surface encloses a volume containing a net charge. Hence it appears from experimental evidence that there does not exist a magnetic charge upon which magnetic field lines may originate or terminate, in the way that electric field lines originate and terminate on electrical charge.

### Faraday's Law and Flux Linkages

If an observer "sees" a changing electric field caused by a time-varying distribution of charges, he will also "see" a magnetic field due to the relative motion of these charges in the process of changing their distribution. More generally, it may be shown that any changing electrical field has associated with it a certain magnetic field. It is not surprising, then, that the converse is also true. In fact, very early experiments with electricity showed that if a loop of wire containing a small gap (as shown in Figure 2.19) were placed in a changing magnetic field, a voltage would appear across the terminals, and

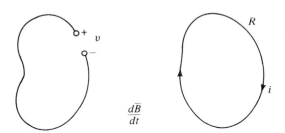

**Figure 2.19** Conductors in a Changing Magnetic Field

if the wire formed a continuous loop, a current would flow around that loop. Both of these phenomena indicate that there is a certain EMF or $\oint \bar{E} \cdot \overline{dl}$ induced by the changing magnetic field. The value of this induced EMF is given by *Faraday's law*, which states that

$$\oint \bar{E} \cdot \overline{dl} = -\frac{d\phi}{dt}, \qquad (2.51)$$

where $\phi$ is the magnetic flux through any surface bounded by the contour around which the line integral is taken (in a right-handed sense). In fact, the physical presence of the wire is not necessary; the contour may simply be an imaginary one in free space. If a wire is present and forms a closed loop of resistance $R$ ohms, then a current of

$$i = \frac{v}{R} = -\frac{1}{R} \oint \overline{E} \cdot \overline{dl} = \frac{1}{R} \frac{d\phi}{dt} \tag{2.52}$$

will flow.

Suppose now that a coil of $N$ turns is constructed, as indicated in Figure 2.20. We will assume that the dimensions of the coil are such that its perimeter may be approximated by a single contour. If we integrate the electric

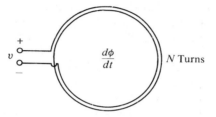

**Figure 2.20**   Multiturn Coil in a Changing Magnetic Field

field along the wire from terminal to terminal to evaluate $v$, we traverse this perimeter contour $N$ times, so the total voltage induced terminal to terminal is

$$v = -N \oint \overline{E} \cdot \overline{dl} = N \frac{d\phi}{dt} = \frac{d\psi}{dt}. \tag{2.53}$$

where $\psi = N\phi$ weber-turns. For obvious reasons $\psi$ is termed *flux linkage*. For a coil of more complicated shape, such as a long solenoid or a toroid, the path in space defined by the wire is more complicated than that of Figure 2.20, but the flux linkages may always be thought of as being defined by equation 2.53; i.e., the induced voltage around the entire circuit is equal to the time rate of change of the flux linkage.

## 2.4   INDUCTIVE PHENOMENA

We have seen (Section 1.3) that a configuration which permits a separation of charges may be characterized by a parameter called capitance, which depends on the configuration and the medium. In an analogous fashion, a configuration that allows a circuital flow of current may be characterized by a parameter called inductance, which also depends on the configuration and the medium. The circuit may consist of a single turn, as in the case of a transmission line, or it may consist of many turns, as in a motor winding or solenoid coil. In any case a flow of current in the circuit will cause a magnetic field and a certain flux linkage. The constant of proportionality between the flux linkage and the current is called the *inductance* of the circuit; i.e.,

$$L \text{ (inductance)} \triangleq \frac{\psi}{i} \qquad \text{weber-turns/ampere, or henrys.} \qquad (2.54)$$

It should be reemphasized that this is a parameter which does not depend upon either the current or the flux linkages per se but only upon their ratio, which in turn depends on the configuration of the circuit and the medium in which the magnetic field exists. Also, this parameter will be present, whether it is desired or not, in any circuit in which a current flows. A device, usually consisting of a number of turns of wire, possibly wound on a core of high permeability to increase the flux linkage, which is built for the specific purpose of obtaining inductance, is called an *inductor*. An inductor may also be defined as a device especially designed to allow energy to be stored in a magnetic field—like capacitors, inductors are energy-storage devices, and an expression for this stored energy will be developed in Chapter 3. Just as in the case of unwanted capacitance, inductance will always be present in any circuit and may be detrimental to the operation for which the circuit was designed. Such unwanted parameters (capacitance or inductance) are often called *stray* or *parasitic parameters*. Another kind of inductance, mutual inductance, will be discussed in Chapter 3.

### Example 2.4

Consider a coaxial cable, as shown in Figure 2.21, where current flows in one direction through the central conductor and returns through the outer conductor. We wish to calculate the inductance per unit length of this configuration. It will be assumed that the permeability everywhere is $\mu_0$ (approximately true for copper conductors).

We must calculate the $\bar{B}$ field in four regions, using Ampère's circuital law.

*Region 1: $r \leq r_1$.*

$$2\pi r B = \mu_0 I \frac{\pi r^2}{\pi r_1^2}, \qquad B = \frac{\mu_0 I r}{2\pi r_1^2}. \qquad (2.55)$$

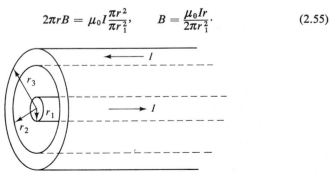

**Figure 2.21**   Coaxial Cable

If we visualize a plane of unit length which contains the axis of our cable, it is clear that in region 1 the flux linkage *per unit length* is

$$\psi_1 = \int_0^{r_1} B \frac{r^2}{r_1^2} \, dr = \int_0^{r_1} \frac{\mu_0 I}{2\pi r_1^4} r^3 \, dr = \frac{\mu_0 I}{8\pi}. \tag{2.56}$$

(*Note:* $r^2/r_1^2$ gives the proportion of the total current linked by a circle of radius $r$.)

*Region 2:* $r_1 < r < r_2$.

$$2\pi r B = \mu_0 I, \qquad B = \frac{\mu_0 I}{2\pi r}. \tag{2.57}$$

In this region, any circle of radius $r$ links all the current $I$; therefore, per unit length,

$$\psi_2 = \int_{r_1}^{r_2} B \, dr = \frac{\mu_0 I}{2\pi} \int_{r_1}^{r_2} \frac{dr}{r} = \frac{\mu_0 I}{2\pi} \ln \frac{r_2}{r_1}. \tag{2.58}$$

*Region 3:* $r_2 < r < r_3$.

The net current linked by a circle of radius $r$ in this region is that carried by the center conductor ($I$), less that proportion of $I$ flowing inside the circle in the outer conductor. Hence in this region

$$2\pi r B = \mu_0 \left( I - I \frac{r^2 - r_2^2}{r_3^2 - r_2^2} \right),$$

or

$$B = \frac{\mu_0 I}{2\pi r} \left( \frac{r_3^2 - r^2}{r_3^2 - r_2^2} \right). \tag{2.59}$$

The flux linkage per unit length in this region is therefore

$$\begin{aligned}
\psi_3 &= \int_{r_2}^{r_3} B \left( 1 - \frac{r^2 - r_2^2}{r_3^2 - r_2^2} \right) dr = \int_{r_2}^{r_3} B \frac{r_3^2 - r^2}{r_3^2 - r_2^2} \, dr \\
&= \frac{\mu_0 I}{2\pi (r_3^2 - r_2^2)^2} \int_{r_2}^{r_3} \frac{(r_3^2 - r^2)^2}{r} \, dr \\
&= \frac{\mu_0 I}{2\pi} \left[ \frac{r_3^4}{(r_3^2 - r_2^2)^2} \ln \frac{r_3}{r_2} - \frac{3r_3^2 - r_2^2}{4(r_3^2 - r_2^2)} \right]. \tag{2.60}
\end{aligned}$$

*Region 4:* $r > r_3$.

From Ampère's circuital law, $B = 0$ and $\psi_4 = 0$.

The total flux linkages per ampere per unit length of the cable is thus

$$\begin{aligned}
L &= \frac{\psi_1 + \psi_2 + \psi_3 + \psi_4}{I} \\
&= \mu_0 \left\{ \frac{1}{8\pi} + \frac{1}{2\pi} \ln \frac{r_2}{r_1} + \frac{1}{2\pi} \left[ \frac{r_3^4}{(r_3^2 - r_2^2)^2} \ln \frac{r_3}{r_2} - \frac{3r_3^2 - r_2^2}{4(r_3^2 - r_2^2)} \right] \right\}. \tag{2.61}
\end{aligned}$$

As is evident from Example 2.4, the calculation of the inductance even for relatively simple configurations tends to be more difficult than the calculation of the capacitance of simple configurations. (Compare, for example, the difficulties in computing the inductance and capacitance per unit length for the coaxial cable; Problem 1.20.) For inductors made up of wound coils of various configurations, an exact calculation of the inductance is usually impractical, although approximate formulas are known in many cases.

### Magnetic Materials and Magnetization

In our discussion of capacitance, we noted that the capacitance of a configuration could be increased by using a medium that polarizes under the influence of an applied electric field. An analogous situation exists in the case of the magnetic field and inductance, but like most magnetic field phenomena, it is somewhat more complicated than electrical polarization.

We have seen that a point charge in motion appears as a finite current element; it is not surprising, then, that an electron in orbit around the nucleus of an atom should have the same effect as a small loop of circulating current. This microscopic current loop (Figure 2.22) has an associated magnetic field and is termed a *magnetic dipole*.

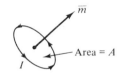

**Figure 2.22**   Magnetic Dipole

If placed in an external magnetic field, this dipole may be shown to experience a torque proportional to the current, the area of the loop, and the sine of the angle between the normal to the plane of the loop and the local $\bar{B}$ vector of the externally applied field. It is therefore useful to define a dipole moment vector, $\bar{m}$, whose magnitude is

$$m = IA. \tag{2.26}$$

In a single atom, there will normally be a number of such dipole moments (one from each orbiting electron) with different spatial orientations. There will also be additional dipole moments, of smaller amplitude, due to the spin of the electrons on their own axes. (These electron spin moments are found always to be an integer multiple of a certain smallest value, called the *Bohr magneton*.)

### Diamagnetic Materials

In certain materials, in the absence of an externally applied magnetic field, the individual atoms show no net dipole moment; i.e., the individual orbital and spin moments of the electrons effectively cancel. In the process of applying a magnetic field, one must establish a certain $d\bar{B}/dt$, which will in turn produce an $\bar{E}$ field, such that around any closed contour Faraday's law will be satisfied. It may be shown that this $\bar{E}$ field will act on the orbiting electrons in such a way that the orbits tend to orient themselves so that the net dipole moment of the atom is such that its field subtracts from the external field. Hence the net $\bar{B}$ field for a certain applied current is decreased, and $\mu < \mu_0$, but only slightly. Copper, for example, is diamagnetic, with

$$\frac{\mu}{\mu_0} = 0.9999906.$$

### Paramagnetic Materials

Paramagnetism occurs in materials whose atoms, in the absence of any applied magnetic field, show a net permanent magnetic moment. If an external field is applied, the primary effect is that these atomic dipoles tend to align themselves with the applied $\bar{B}$ field (because of the previously discussed torque which they experience), thus increasing the net $\bar{B}$ field for a given applied current. Thus, in this case, $\mu > \mu_0$, but, again, only slightly. Aluminum, for example, is paramagnetic, with

$$\frac{\mu}{\mu_0} = 1.0000214.$$

### Ferromagnetic Materials

In an important class of materials, called *ferromagnetic*, the atoms not only display permanent moments (paramagnetism), but clusters of many atoms are found with their magnetic moments already perfectly aligned. Such a cluster is called a magnetic *domain*, which, although microscopic in size, may still contain $10^{20}$ atoms or more. Under the influence of an applied field, the domains change in size and orientation with respect to each other, and the resultant net increase in the $\bar{B}$ field can be large, so $\mu \gg \mu_0$. However, this effect, unlike diamagnetism or paramagnetism, tends to be highly nonlinear. For small values of magnetizing current, the domains are not reoriented to any large degree, while for large values of current, the domains are nearly all lined up with the field, so the effect ceases. If the current is then made zero, the domains do not return completely to their original configuration, so the $\bar{B}$ field does not return to zero for the macroscopic piece of material—this lagging effect of the domain orientation is called *hysteresis* and is illustrated in Figure 2.23. Since the relative permeability of a ferromagnetic material, such as soft iron, may be several thousand over the steepest portion of this curve, a logical way of greatly increasing the inductance of a coil is to wind the coil on a core of such material, as illustrated in Figure 2.24.

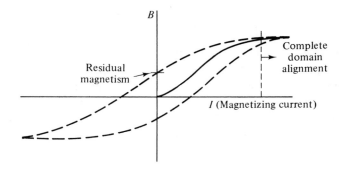

**Figure 2.23**  Magnetizing and Hysteresis Curve

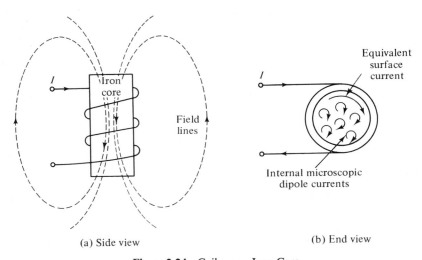

(a) Side view

(b) End view

**Figure 2.24**  Coil on an Iron Core

As illustrated in Figure 2.24(b), the internal atomic circulating currents of the dipoles tend to cancel, but this cancellation does not take place at the boundary of the core; hence the macroscopic effect of the aligned dipoles is the same as an equivalent surface current flowing around the core in the same direction as the winding current, thus increasing the magnetic field. To further increase the inductance, we may let the iron core close on itself, so the field lines (which must form closed contours) lie for the most part within the material, as indicated in Figure 2.25. In this configuration, if $\mu \gg \mu_0$, nearly all the magnetic flux may be considered to be within the core, although there will always be some magnetic field in the space surrounding the core, and this is usually called *leakage flux*. Configurations consisting of two or more coils will be discussed in Chapter 4.

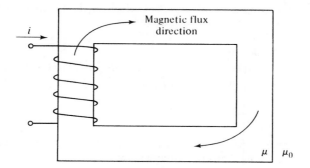

**Figure 2.25** Closed Core

## PROBLEMS

**2.1** An electron beam has a cross-sectional area of 1 mm² and has an electron density of $10^{15}$ electrons/m³. If the velocity of the electrons is 1000 m/s, calculate the beam current.

**2.2** A copper wire with a cross-sectional area of 1 mm² carries a current of 1 A. Calculate the average drift velocity of the free electrons in the wire.

**2.3** The charge that has passed through a cross section of a wire is given by

$$q(t) = \sin \omega t \qquad \text{coulombs.}$$

Find the current $i(t)$ in the wire.

**2.4** The current in a wire is specified as

$$i(t) = \sin \omega t \qquad \text{amperes.}$$

At any time $t$, find the charge that has passed through a cross section of the wire. Is there a single unique answer to this question? Explain carefully.

**2.5** Explain the terms "through variable" and "across variable" as applied to voltage and current.

**2.6** In a wire carrying a steady current, the current density vector constitutes a uniform field. Show that the total current, which might be obtained by integrating $\bar{j} \cdot \overline{ds}$ over any cross-sectional surface, is independent of the orientation of the surface (i.e., the angle that it makes with the wire).

**2.7** If all the free charge in 1 cm³ of copper were placed at a distance of 1 m from a similar charge, find the force (in tons) that would be required to maintain this separation.

**2.8** Using a typical value of $\rho = 1.7 \times 10^{-8}$ ohm-meter, find the resistance of a copper wire 1 m long and 1 mm² in cross-sectional area.

**2.9** An electron is injected with an initial velocity $\bar{u}_0$ into a uniform $\bar{B}$ field. Find the trajectory of the electron if $\bar{u}_0$ is
(a) Parallel to $\bar{B}$.
(b) Perpendicular to $\bar{B}$.

**2.10** By integrating equation 2.40, show that the $\bar{B}$ field at a distance $r$ from a straight wire carrying a current $I$ has a magnitude

$$B = \frac{\mu I}{2\pi r}$$

and discuss its direction. (Contrast the difficulty of integrating equation 2.40 with that of using Ampère's circuital law.)

**2.11** An electron moves parallel to, and at a distance of 1 cm from, a long, straight wire carrying a current of 10 A. If the electron speed is $10^4$ m/s, find the force exerted on it.

**2.12** Two electron beams, each having a linear charge density of $\rho$ coulombs/meter, move parallel to each other, at a distance of 1 m, and both have the same velocity $\bar{u}$ with respect to an observer. Find the ratio of the magnetic force per unit length to the electrostatic force per unit length as seen by the observer. (Compare with Example 2.2.) What force will be observed if the observer also moves at the velocity $\bar{u}$?

**2.13** Figure 2.26 shows the charge through a cross section of wire as a function of time. Sketch the current $i(t)$, with enough information to specify it completely.

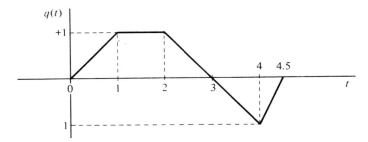

**Figure 2.26** Charge Waveform

**2.14** If the function sketched in Figure 2.26 represents the current in a wire, sketch the charge transported, $q(t)$, with enough information to specify the function completely.

**2.15** Two parallel wires, separated by a distance of 1 m, each carry a current of 1 A, in opposite directions. The wires are in free space. Using Ampère's circuital law (and superposition), calculate the $\bar{B}$ field between the two wires, in a plane containing the wires.

**2.16** If the wires of Problem 2.15 have a radius of 1 mm, find the approximate inductance of the configuration per unit length, neglecting the flux linkage due to the $\bar{B}$ field within the wires themselves. Why is it necessary to assume a finite (nonzero) radius for the wires?

**2.17** Repeat Problem 2.16 without neglecting the flux linkage due to the $\bar{B}$ field within the wires.

**2.18** A circular loop of wire of radius $R$ carries a current of $I$ amperes. Find the $\bar{B}$ field at all points along the axis of the loop.

**2.19** A circular loop of wire of radius $R$ is placed in a uniform $\bar{B}$ field such that the field is perpendicular to the plane of the loop. If the $\bar{B}$ field varies with time according to

$$\bar{B}(t) = B \sin \omega t \, \bar{i}_\perp,$$

find the induced EMF around the loop.

**2.20** A loop of wire of resistance $R$ ohms and area $A$ rotates with a constant angular velocity $\omega$ rad/s in a uniform $\bar{B}$ field, as indicated in Figure 2.27. Calculate the induced current in the loop.

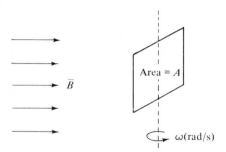

**Figure 2.27** Rotating Loop in Magnetic Field

**2.21** A D'Arsonval movement, used in many different types of indicating instruments, is illustrated in Figure 2.28. The magnet pole faces are shaped so that

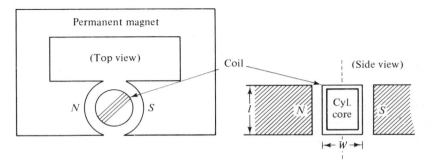

**Figure 2.28** D'Arsonval Movement

the $\bar{B}$ field in the air gap is approximately cylindrically radial. If the coil consists of $N$ turns and carries a current of $I$ amperes, find the torque exerted on the coil. If the coil is restrained by a spring having a spring constant of $K$ newton-meters/radian, what will be the equilibrium position of the coil?

**2.22** An iron rod has a square cross section, 1 cm $\times$ 1 cm, and a permeability of $\mu$. The rod is bent to form a circular ring, of inner radius 10 cm and outer radius 11 cm, and the ends are welded together. Wire is wound toroidally around the ring to form a coil of $N$ turns, and a current of $I$ amperes is passed through this coil.
Calculate:
(a) The $\bar{B}$ field at all points within the iron (use Ampère's circuital law).
(b) The flux of $\bar{B}$ through the cross section of the iron ring.
(c) The inductance of the configuration.

**2.23** A conducting rod of length $L$ rotates at $\omega$ radians/second about an axis perpendicular to the axis of the rod. If the rotating rod is situated in a uniform $\bar{B}$ field parallel to the axis of rotation, find the force caused by the magnetic field on each electron in the rod. What happens to the free charge in the rod?

**2.24** A conducting bar slides on two parallel conducting tracks toward a fixed bar, as indicated in Figure 2.29. Both the bars and the tracks have a resistance of $R$ ohms/meter. A uniform $\bar{B}$ field links the configuration (upward and normal to the plane) as shown. Find the magnitude and direction of the current that flows as the sliding bar is moved with a velocity $\bar{u}$ as shown. How much power is required to move the bar at this velocity? How much force is required?

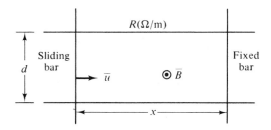

**Figure 2.29** Configuration for Problem 2.24

# 3

# *Linear Circuit Elements and Networks*

### 3.1 PASSIVE CIRCUIT ELEMENTS: TERMINAL CHARACTERISTICS

Circuit elements may be classified as being either passive or active. *Active elements* (batteries, generators, transistors, etc.) involve external energy sources and provide the driving or forcing functions in circuits or networks. *Passive elements*, on the contrary, do not involve external energy sources but are characterized by predominantly exhibiting one of the three basic electrical properties discussed in the preceding chapters: resistance, inductance, or capacitance. This set of passive elements consists of resistors, inductors, capacitors, and coupled (or mutual) inductors.

### *Resistors*

Physical resistors may have resistance values ranging from fractions of an ohm to many megohms (1 megohm = $10^6$ ohms). Typically, resistors consist of metal strips or metal wire in various configurations, films of metal or carbon deposited on glass or ceramic rods, powdered carbon mixed with an inert material and a bonding agent, or, in modern integrated circuits, a thin filament of conducting or semiconducting material deposited on other semiconductors in a monolithic structure. Such a device, built essentially to display the property of resistance, is indicated on circuit diagrams by the symbol of Figure 3.1(a), and, with regard to its two terminals, displays the voltage–current relationship of equation 2.23 (Ohm's law):

$$v(t) = Ri(t). \qquad (2.23)$$

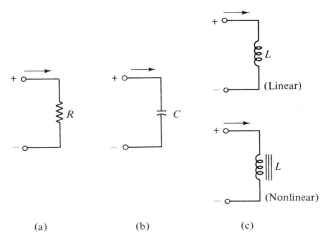

(a)                     (b)                     (c)

**Figure 3.1**  Symbols for Passive Circuit Elements

The function of a resistor, basically, is to dissipate energy in the form of heat, as discussed in Chapter 2, and the power relationship at the terminals is given by equation 2.29:

$$p(t) = v(t)i(t) = Ri^2(t) = \frac{v^2(t)}{R}.$$  (2.29)

### Capacitors

A device that displays predominantly the property of capacitance is called a *capacitor* ("condenser" is an equivalent, although somewhat obsolete, term) and is indicated in a circuit diagram by the symbol of Figure 3.1(b). Typical values of capacitors used in circuits range from a few picofarads (1 picofarad $= 10^{-12}$ farad), to several hundred microfarads (1 microfarad $= 10^{-6}$ farad), and physically consist of a wide variety of separated-conductor configurations in various dielectric materials. (In integrated circuits capacitors may be easily constructed by successive depositions of conducting and dielectric films.) The terminal voltage–current characteristic of a capacitor may be easily derived from the defining equation of capacitance, equation 1.36, and equation 2.4, which relates charge and current; i.e.,

$$q(t) = \int_{-\infty}^{t} i(\tau)\, d\tau = Cv(t).$$  (3.1)

Thus,

$$v(t) = \frac{1}{C} \int_{-\infty}^{t} i(\tau)\, d\tau,$$  (3.2)

or conversely,

$$i(t) = C\frac{dv}{dt}. \tag{3.3}$$

If the capacitance itself is time varying, then equation 3.3 requires modification. In this case

$$q(t) = \int_{-\infty}^{t} i(\tau)\, d\tau = C(t)v(t), \tag{3.4}$$

$$v(t) = \frac{1}{C(t)}\int_{-\infty}^{t} i(\tau)\, d\tau, \tag{3.5}$$

and

$$i(t) = \frac{d}{dt}[C(t)v(t)]$$

$$= C(t)\frac{dv}{dt} + v(t)\frac{dC}{dt}. \tag{3.6}$$

Basically, the capacitor is an *energy storage device*; all of the terminal power $v(t)\, i(t)$ goes into stored energy, and the instantaneous value of this stored energy for a linear capacitor may be obtained from equation 1.44; i.e.,

$$w(t) = \frac{1}{2}Cv^2(t). \tag{3.7}$$

### *Inductors*

Physical inductors consist typically of coils of wire of various configurations (solenoids, toroids, etc.), either with air cores or cores of ferromagnetic material. Coils on nonmagnetic or air cores yield nearly linear inductors (the meaning of which will be discussed shortly), while coils wound on ferromagnetic cores yield inductors that can only be treated as linear over small variations of current. Over large current variations, one must take into account not only the nonlinearity of the magnetization curve (Figure 2.23), but the hysteresis phenomena as well, making exact analysis difficult. Linear and nonlinear (ferromagnetic core) inductors are usually indicated in circuits by the symbols of Figure 3.1(c).

For a linear inductor, i.e., one in which the flux linkages are directly proportional to the current, the terminal voltage–current characteristic may be found by using equations 2.53 and 2.54; i.e.,

$$v(t) = \frac{d\psi}{dt} = L\frac{di}{dt}, \tag{3.8}$$

or conversely,

$$i(t) = \frac{1}{L} \int_{-\infty}^{t} v(\tau) \, d\tau. \tag{3.9}$$

If the inductor itself is time-varying, we must write

$$v(t) = \frac{d\psi}{dt} = \frac{d}{dt}[L(t)i(t)]$$

$$= L(t)\frac{di}{dt} + i(t)\frac{dL}{dt}, \tag{3.10}$$

and conversely,

$$i(t) = \frac{1}{L(t)} \int_{-\infty}^{t} v(\tau) \, d\tau. \tag{3.11}$$

To calculate the energy stored in the inductor, we may write

$$w(t) = \int_{-\infty}^{t} p(\tau) \, d\tau = \int_{-\infty}^{t} i(\tau)v(\tau) \, d\tau$$

$$= L \int_{-\infty}^{t} i(\tau)\frac{di(\tau)}{d\tau} \, d\tau = L \int_{i(-\infty)}^{i(t)} i(\tau) \, di(\tau). \tag{3.12}$$

Assuming that $i(-\infty) = 0$,

$$w(t) = \frac{1}{2}Li^2(t). \tag{3.13}$$

### Idealization of the Element Models

We may now summarize the voltage–current terminal characteristics of linear time-invariant resistors, inductors, and capacitors, as follows:

$$R: \quad v(t) = Ri(t) \qquad\qquad i(t) = \frac{1}{R}v(t)$$

$$L: \quad v(t) = L\frac{di}{dt} \qquad\qquad i(t) = \frac{1}{L} \int_{-\infty}^{t} v(\tau) \, d\tau$$

$$C: \quad v(t) = \frac{1}{C} \int_{-\infty}^{t} i(\tau) \, d\tau \qquad i(t) = C\frac{dv}{dt}$$

In the derivation of these terminal characteristics, we assumed that the elements were pure, and this is the assumption that is made when the symbols of Figure 3.1 are employed in circuit diagrams. As we have emphasized previously, however, this is only an approximation; each of the physical elements will exhibit all the parameters of resistance, inductance, and capacitance to some extent. In a coil, for example, the wire will have some resistance, and capacitance will exist between adjacent (and nonadjacent) turns. A capacitor will have some resistance and inductance associated with its lead wires, and a certain conductivity associated with the dielectric (unless it is vacuum). A resistor carries current and will therefore have some associated inductance, and since different portions of the resistor are at different potentials, a capacitive effect is present. Hence, to adequately model a physical coil, capacitor, or resistor, particularly when the rates of change (derivatives) of the voltage and current are large, it may be necessary to use a combination of ideal elements. Thus, considering actual physical configurations, real resistors, inductors, and capacitors might logically be represented by ideal elements using the equivalent circuits of Figure 3.2.

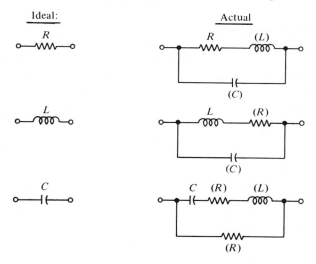

**Figure 3.2** Possible Representations of Physical Circuit Elements

[Even these equivalent-circuit representations may become inadequate if the derivatives become too large—as rates of change increase, the physical device eventually must be treated as a distributed device (involving partial derivatives) rather than a two-terminal lumped device.] The existence of these parasitic elements of course does not mean that network analysis based upon ideal elements is not valid; it simply means that some care must be taken in relating the idealized circuit model to the real-world network.

### Mechanical Analogies of R, L, and C

As an aid to understanding the behavior of resistors, inductors, and capacitors, it is helpful to consider certain analogous mechanical elements:

masses, springs, and viscous damping elements (dash-pots). Denoting mass by $M$ (kilograms), spring constants by $K$ (newtons/meter), and viscous damping constants by $B$ (newtons/meter/second), and letting $F$ and $V$ denote force and velocity, respectively, we have the relationships:
For a mass,

$$F(t) = M\frac{dV}{dt}, \tag{3.14}$$

for a spring,

$$F(t) = K\int_{-\infty}^{t} V(\tau)\,d\tau, \tag{3.15}$$

and for a viscous damper,

$$F(t) = BV(t). \tag{3.16}$$

Thus, if we consider voltage to be analogous to force, $(v \sim F)$, and current analogous to velocity, $(i \sim V)$, it follows that $L \sim M$, $C \sim 1/K$, and $R \sim B$. Thus in any circuit composed of $R$'s, $L$'s, and $C$'s, if we consider these elements to be replaced by their analogous mechanical elements, then all the force–velocity relationships in the mechanical network will be identical to the voltage–current relationships in the electrical network. (We may also establish the analogy in another way—with $v \sim V$ and $i \sim F$; these analogies will be elaborated on in the problems of Chapter 3.)

### Example 3.1

If a mass is subjected to a constant force, the velocity of that mass increases linearly with time. If a constant voltage is applied to the terminals of an inductor, the current in that inductor increases linearly with time. A mass subject to no force travels at a constant velocity; an inductor carrying a constant current has no potential difference between its terminals.

A spring subjected to a constant force exhibits a constant deflection—i.e., no velocity difference between its ends. A capacitor with a constant applied voltage passes no current. A constant velocity between the ends of a spring means a linearly increasing force on the spring; a constant current through a capacitor implies a linearly increasing voltage across that capacitor.

### Mutual Inductance

A common network configuration involves magnetic flux from one coil linking another coil (or coils), and a rate of change of this flux will then induce an EMF in both coils (or all of the coils which it links). For example, consider the transformer configuration of Figure 3.3(a). Assuming that all the flux is in the core material, and using the right-hand rule for the flux direction, it is clear that the two currents, $i_1$ and $i_2$, produce fluxes that add (within the assumption of linearity). On the other hand, if one of the winding directions

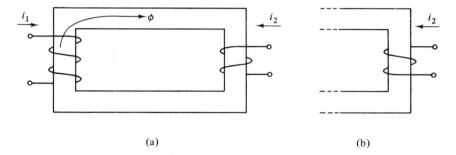

(a)                                                    (b)

**Figure 3.3**   Two Coupled Coils

is reversed, as in Figure 3.3(b), the fluxes will subtract. On a circuit diagram, whether the fluxes add or subtract is often indicated by the *dot convention*, and the same convention is often used to mark actual transformers in order to indicate winding directions. The dot convention may be stated as follows: A dot is placed near one of the terminals of each winding, so currents flowing into (or out of) both dotted terminals produce fluxes that add. Thus, in a circuit diagram, coupled coils may be indicated as in Figure 3.4; with cur-

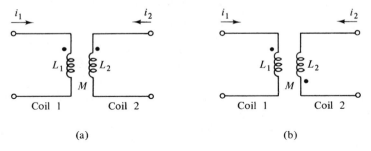

(a)                                                    (b)

**Figure 3.4**   Circuit Diagram for Two Coupled Coils

rents flowing in the indicated reference directions, in (a) the fluxes add and in (b) they subtract. Referring to Figure 3.4(a), we will define the following:

$$\psi_{11}: \quad \text{flux linkage of coil 1 due to } i_1$$
$$\psi_{12}: \quad \text{flux linkage of coil 1 due to } i_2$$
$$\psi_{21}: \quad \text{flux linkage of coil 2 due to } i_1$$
$$\psi_{22}: \quad \text{flux linkage of coil 2 due to } i_2$$
$$\psi_1 \ : \quad \text{total flux linkage of coil 1}$$
$$\psi_2 \ : \quad \text{total flux linkage of coil 2.}$$

Then

$$v_1 = \frac{d\psi_1}{dt} = \frac{d\psi_{11}}{dt} + \frac{d\psi_{12}}{dt} = L_1 \frac{di_1}{dt} + M_{12} \frac{di_2}{dt}$$

$$v_2 = \frac{d\psi_2}{dt} = \frac{d\psi_{21}}{dt} + \frac{d\psi_{22}}{dt} = M_{21} \frac{di_1}{dt} + L_2 \frac{di_2}{dt}, \quad (3.17)$$

where

$$M_{12} \triangleq \frac{\psi_{12}}{i_2} \quad \text{and} \quad M_{21} \triangleq \frac{\psi_{21}}{i_1} \quad \text{(henrys).} \quad (3.18)$$

$L_1$ and $L_2$ are called the *self-inductances* of the two coils, and $M_{12}$ and $M_{21}$ are called *mutual inductances*. (We will show shortly that $M_{12} = M_{21}$.)

If the fluxes associated with the indicated currents subtract rather than add, then the mutual inductance terms in equations 3.17 will be negative (i.e., $M_{12}$ and $M_{21}$ will be negative quantities). Hence the dot convention may be thought of as a specification of the sign of the mutual inductance. (If more than two coils are involved, the sign of the mutual inductance for each pair must be specified; hence for three coils, three different sets of "dots," of distinguishable shapes, must be used. An example appears in the problems.)

We will now calculate the energy stored in the coupled coils of Figure 3.4(a). If $i_2$ is held at zero value (open-circuited), then the energy required to bring the current in coil 1 from zero (say at time $-\infty$) to the value $i_1$ at time $t_1$ is

$$w_1 = \int_{-\infty}^{t} i_1 v_1 \, d\tau = L_1 \int_{-\infty}^{t} i_1 \frac{di_1}{d\tau} \, d\tau = L_1 \int_{0}^{i_1} i_1 \, di_1 = \tfrac{1}{2}L_1 i_1^2, \quad (3.19)$$

just as if coil 2 were not present at all. We will now hold this current $i_1$ in coil 1 fixed, and bring the current in coil 2 up to the value $i_2$. During this process, since $di_1/dt = 0$, $v_1 = M_{12}(di_2/dt)$ and $v_2 = L_2(di_2/dt)$. Therefore, the work done is

$$w_2 = \int_{-\infty}^{t} i_1 v_1 \, d\tau + \int_{-\infty}^{t} i_2 v_2 \, d\tau$$

$$= i_1 \int_{-\infty}^{t} M_{12} \frac{di_2}{d\tau} \, d\tau + \int_{-\infty}^{t} L_2 i_2 \frac{di_2}{d\tau} \, d\tau$$

$$= M_{12} i_1 \int_{0}^{i_2} di_2 + L_2 \int_{0}^{i_2} i_2 \, di_2$$

$$= M_{12} i_1 i_2 + \tfrac{1}{2}L_2 i_2^2. \quad (3.20)$$

Therefore, the total stored energy is

$$w = w_1 + w_2 = \frac{1}{2}L_1 i_1^2 + \frac{1}{2}L_2 i_2^2 + M_{12} i_1 i_2. \quad (3.21)$$

Had we brought up the two currents in the reverse order, we would have obtained the same energy as given in equation 3.21, but with $M_{12}$ replaced by $M_{21}$. Since the total magnetic field must be the same in either case, it follows that $M_{21} = M_{12}$. That is, there is only one value of mutual inductance $(M)$ associated with any pair of coils.

## 3.2 ACTIVE CIRCUIT ELEMENTS: SOURCES

### Voltage and Current Sources

A *source* is a network element that is capable of supplying energy to a network, energy that is either stored or dissipated in the passive network elements. Typical physical sources include batteries, electric generators of many diverse types and sizes, and vacuum-tube or transistor circuits. (The source itself may involve a complicated interconnection of circuits and devices, but only its terminal characteristics will be of importance when it is used as a driving element in another network.) For the purpose of network analysis, it is useful to define idealized source elements, just as we defined idealized resistors, inductors, and capacitors. We will therefore define an ideal voltage source and an ideal current source as follows.

An *ideal voltage source* is an element that maintains the voltage between its terminals (as either a specified constant value or specified time function) regardless of what is connected to these terminals.

An *ideal current source* is an element that maintains a current flow between its terminals (as either a specified constant value or specified time function) regardless of what is connected to these terminals.

The usual symbols used to indicate these ideal sources are indicated in Figure 3.5. An *open circuit* may be considered a current source of zero value (by definition, no current can flow between open-circuited terminals). Similarly, a *short circuit* may be considered a voltage source of zero value (by definition, a short circuit cannot have any nonzero potential difference between its terminals). From the source definitions, it follows that two ideal voltage sources can be connected in parallel between two terminals only if they have identical values (since only one potential difference can exist

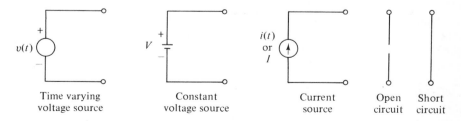

**Figure 3.5** Symbols Indicating Ideal Sources

between two points). Similarly, two ideal current sources can be connected in series only if both are identical (since along any one path in a circuit we can have only one current). For these reasons it is "illegal" to have an ideal voltage source of nonzero value with an ideal short circuit across its terminals, or an ideal nonzero current source that is open-circuited. On the other hand, an ideal voltage source that is open-circuited is idle, since the current it supplies and therefore the power are zero. Similarly, an ideal current source is idle when short-circuited, since the voltage across its terminals and therefore the power supplied are zero. The terminal characteristics of ideal voltage and current sources are shown in Figure 3.6.

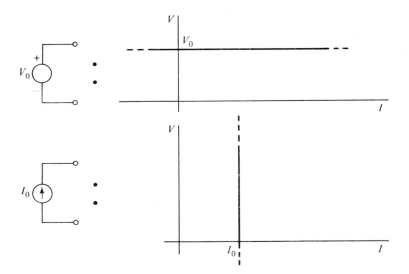

**Figure 3.6** *V-I* Characteristics of Ideal Constant Sources

### Example 3.2

Figure 3.7 shows a voltage source, a current source, and a load resistor, all connected in parallel. The voltage source, by definition, maintains a potential of $V$ volts across both the current source and the resistor, regardless of their values. Similarly, the current source causes 1 A to flow in the middle branch

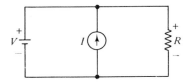

**Figure 3.7** Network Containing Ideal Sources

of the circuit, regardless of the values of either the voltage source or the resistor.

### Dependent Sources

If the terminal voltage of a voltage source or the current supplied by a current source are determined by factors external to the circuit being analyzed, the source is called an independent source. Often, however, the source is of such a nature (a vacuum-tube or transistor, for example) that its terminal voltage or current depends upon some other voltage or current in the circuit being analyzed; in this case, the source is termed a dependent source. The same symbols are used for dependent and independent sources in circuit diagrams, but the dependent source is ordinarily labeled to indicate clearly what voltage or current in the circuit the source depends upon, and what function of that voltage or current is produced by the source. An example is discussed in the next section.

### Junction Transistor[1]

In modern electrical engineering, probably the most important example of a dependent source is the transistor, which may be used in either a linear or a highly nonlinear mode. For this reason, and because later we will want to examine the behavior of circuits involving transistor amplifiers, a very brief qualitative description of junction transistor operation and equivalent circuits is presented.

The most widely used dependent element in electronic amplifiers is the junction transistor, invented by William Shockley at Bell Telephone Laboratories in 1948. (Another type of transistor, the field-effect transistor, or FET, is not so commonly used in the linear fashion of interest to us here, and will not be discussed.) Of course, a detailed and quantitative study of transistor physics and terminal characteristics is beyond the scope and intent of this text and rightfully belongs in a course in electronics. Just as in the case of the passive circuit elements, however, some descriptive insight into the physics of the device is necessary before one can feel at home with the idealized circuit model.

The junction transistor is a three-terminal device and consists of two junctions, in either an *n-p-n* or a *p-n-p* sequence. The *n-p-n* sequence is diagrammed, along with two biasing batteries, in Figure 3.8(a). (It should be pointed out that this figure is not intended to convey an accurate physical picture of a junction transistor—in practice, for example, the junctions are often formed by a vacuum deposition of very thin layers of material.) In this

---

[1]The purpose of this section is to convey to the reader a very rudimentary picture of transistor operation. The discussion appears here because of its logical connection to a discussion of sources. The reader may skip this section if he so desires, or at least defer it until after the discussion of Kirchhoff's laws later in the chapter.

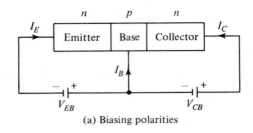

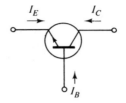

(b) Circuit symbol

**Figure 3.8** *n-p-n* Transistor

discussion we shall consider only the *n-p-n* configuration; with the *p-n-p* configuration, the polarities of the biasing batteries are reversed, and the direction of the arrow in the circuit symbol [Figure 3.8(b)] is also reversed.

As indicated in Figure 3.8(a), the emitter–base junction is forward-biased, while the base–collector junction is reverse-biased (in normal operation). This results in a potential distribution through the transistor as indicated in Figure 3.9. In an actual transistor the base is a very thin layer and is doped (with impurity acceptors) relatively lightly, so that electrons from the emitter which cross the emitter–base boundary nearly all diffuse across the base and are accelerated by the large potential hill into the collector region, where they recombine with holes from the external battery. This

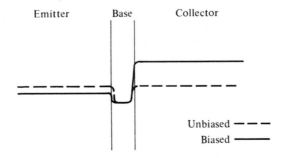

**Figure 3.9** Potential Distribution

electron current, $\alpha I_E$, where $\alpha$ is only slightly less than unity (i.e., about 0.9 to 0.999) constitutes by far the largest part of the collector current, and it is nearly independent of the collector voltage, $V_{CB}$. In addition, there is a small reverse-bias current, due to thermal agitation, across the base–collector junction, so the total collector current may be expressed as

$$I_C = I_{CO} - \alpha I_E,$$

(3.22)

(small)  (large)

where $I_{CO}$ is called the *collector cutoff current* and $\alpha$ is called the *forward current transfer ratio*. The base current is simply the difference between the nearly equal emitter and collector currents; with the current reference directions shown,

$$I_B = -I_E - I_C.$$

(3.23)

(small)

### Common-Base Configuration

The biasing voltages $V_{EB}$ and $V_{CB}$ establish equilibrium or quiescent values of voltages and currents in the transistor. In normal linear operation, however, we are interested in small variations of voltage and current about these equilibrium values, variations caused by the addition of signal sources and load resistors to the circuit. Figure 3.10 indicates an *n-p-n* transistor in the

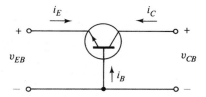

**Figure 3.10**   Common-base Configuration

common-base configuration. The voltages and currents are now indicated by lower case letters rather than capitals as in Figure 3.8, to indicate that these voltages and currents constitute small variations about the equilibrium or quiescent values established by the biasing batteries. Since the emitter–base junction is forward-biased, the *v* vs. *i* characteristic for this junction is essentially that of the first quadrant of Figure 2.12; $v_{CB}$ has little effect on the emitter, as previously noted. This $v_{EB}$ vs. $i_E$ characteristic is indicated in Figure 3.11(a). As previously noted, in equation 3.22, the collector current is essentially that of a reverse-biased junction (third quadrant of Figure 2.12), plus the large component $(-\alpha i_E)$, which is essentially independent of $v_{CB}$. This yields a set of curves, one for each different value of $i_E$, as shown in

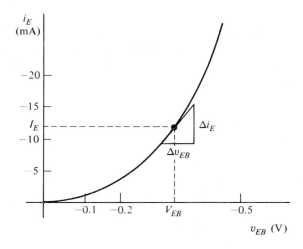

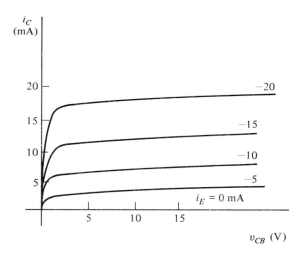

**Figure 3.11** Common-base Characteristics of *n-p-n* Transistor

Figure 3.11(b). (The actual numerical values, of course, vary among different transistors—the numbers indicated are typical.)

In linear operation, the excursions of $v_{EB}$ and $i_E$ about the bias or operating point are small, as indicated by the increments in Figure 3.11(a), and the curve may be approximated by a straight line whose slope indicates the apparent resistance of the emitter–base junction, $R_E$:

$$R_E \triangleq \frac{\Delta v_{EB}}{\Delta i_E} \qquad \text{(dynamic emitter resistance)}. \qquad (3.24)$$

From the collector characteristic, Figure 3.11(b), it is clear that the collector approximates closely a current source, dependent on $i_E$ but nearly independent of $v_{CB}$. Thus for small-increment operation, the transistor characteristics may be approximated by the small-signal equivalent circuit of Figure 3.12. From this equivalent circuit it is easy to see how *voltage amplification* may be

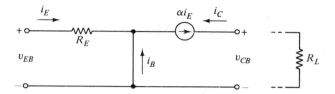

**Figure 3.12**   Small-signal Common-base Equivalent Circuit

achieved. Suppose that a load resistor $R_L$ is connected across the collector–base terminals. Then

$$i_E = \frac{v_{EB}}{R_E},\tag{3.25}$$

$$v_{CB} = -R_L i_C = R_L \alpha i_E = \frac{R_L \alpha v_{EB}}{R_E},\tag{3.26}$$

$$\frac{v_{CB}}{v_{EB}} = \frac{\alpha R_L}{R_E}.\tag{3.27}$$

For example, if $\alpha = 0.9$, $R_E = 50$, and $R_L = 5000$, which represent fairly typical values, then the voltage amplification is

$$\frac{v_{CB}}{v_{EB}} = \frac{0.9 \times 5000}{50} = 90.\tag{3.28}$$

### Common-Base Configuration

Figure 3.13 indicates the *n-p-n* transistor in the "common-emitter" configuration. Again, the voltages and currents indicated by the lower case

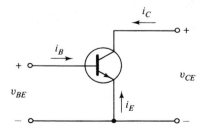

**Figure 3.13**   Common-emitter Configuration

letters are incremental ones; the biasing voltages, indicated in Figure 3.8, are not shown. From equation 3.22, the incremental values $i_C$ and $i_E$ are again approximately related by

$$i_C = -\alpha i_F, \qquad (3.29)$$

and incrementally,

$$i_B = -i_E - i_C, \qquad (3.30)$$

as before. Thus

$$i_C = \alpha(i_B + i_C),$$

or

$$i_C = \frac{\alpha}{1-\alpha} i_B = \beta i_B, \qquad (3.31)$$

where $\beta \triangleq \alpha/(1 - \alpha)$. Also,

$$v_{BE} = -v_{EB} = -R_E i_E \qquad \text{(equation 3.25)} \qquad (3.32)$$

and

$$i_E = -i_B - i_C = -i_B - \beta i_B = -(1 + \beta)i_B. \qquad (3.33)$$

Hence the new input resistance is

$$R_B = \frac{v_{BE}}{i_B} = (1 + \beta)R_E \simeq \beta R_E. \qquad (3.34)$$

Equations 3.31 and 3.34 may be summarized by employing the equivalent (small-signal) equivalent circuit of Figure 3.14 for the common-emitter

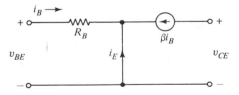

**Figure 3.14**  Small-signal Common-emitter Equivalent Circuit for Idealized Transistor

configuration. Note that in the common-base configuration, no current amplification is possible, since $\alpha < 1$. In the common-emitter circuit, however, current amplification takes place, since $\beta \gg 1$. In addition, by using a large load resistance, we may have voltage amplification as well, and hence the common-emitter configuration is used very extensively in electronic circuits.

We have given a brief introduction to the junction transistor and certain linearized equivalent circuits, because it is probably the most common example of a *dependent source* in electrical engineering. It must again be emphasized that these equivalent circuits represent highly idealized models of actual transistor behavior; in a text on transistor electronics, more refined models, which include some additional elements to account for certain physical phenomena that we have ignored, may be found. Still, the models presented here are useful and valid representations when the amplitudes and rates of change of the signals are not too great, and we will employ these equivalent circuits in examples.

## 3.3 NETWORKS

We have already considered some simple networks and have made use of some of the concepts that we will now formalize for more general networks. The networks that we will consider will consist of interconnections of the various elements already discussed, i.e., ideal resistors, inductors, capacitors, coupled inductors, voltage sources, and current sources.

### Sign Conventions in Networks

In previous examples we have assigned $+$ and $-$ symbols to indicate the polarities of voltages across elements, and arrows to indicate directions of current flow. [In some texts an arrow is also used to indicate voltage polarity; in this text, however, we will continue to use the $(+, -)$ convention.] In discussing polarities and current directions, however, one must distinguish between *reference* polarities and directions and *actual* polarities and directions. The usual convention is as follows: the symbols used in circuit diagrams indicate reference polarities and directions; the actual directions of voltages and currents in the circuit are designated by associating with the reference directions a positive or negative numerical value.

Current flowing through a resistor causes a voltage $v(t) = Ri(t)$ across that resistor, the polarity and current direction being consistent with those indicated in Figure 3.15(a). In an actual circuit application, if the current through the resistor is negative, the voltage will be also; i.e., at any instant of time, the current flows from the positive to the negative terminal of the resistor. For the inductor and capacitor, the reference polarities and directions indicated in Figure 3.15(b) and (c) are consistent with the equations

$$v_L(t) = L\frac{di_L}{dt} \quad \text{and} \quad v_C(t) = \frac{1}{C} \int_{-\infty}^{t} i_C(\tau) \, d\tau.$$

Hence, as far as reference directions are concerned, going through any passive element in the direction of the reference current, we experience a drop of

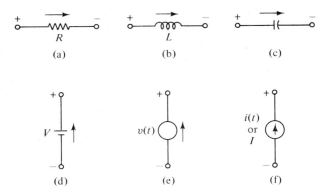

**Figure 3.15** Reference Directions and Polarities

potential or a *voltage drop*. Conversely, if we "go through" any passive element in the direction opposite to that assigned as the current reference, we see a *voltage rise*. (Of course, in terms of actual directions, a negative voltage rise is a voltage drop, and vice versa.) Thus, for a resistor, the actual voltage and current must always have the same sign under this convention, but this is not true for the inductor or capacitor. With the inductor, for example, $di/dt$ may be negative when $i(t)$ is positive, and vice versa, and similarly, with the capacitor, $\int_{-\infty}^{t} i(\tau)\, d\tau$ may be negative when $i(t)$ is positive, and vice versa. While the reference directions may be assigned immediately, the actual directions of voltages and currents must generally be found by solving the network equations.

Figure 3.16 shows a voltage source and a current source connected to a single passive element. If the connecting wires are assumed to be ideal short circuits, then in each of these circuits there can be only one potential difference, and of course there can be only one current; hence the reference polarities and current directions for the sources are most logically assigned as indicated. Thus in going through a source in the direction of the assigned current reference, we experience a voltage rise, and the instantaneous power supplied by the source is $v(t)i(t)$, this power being either dissipated or going

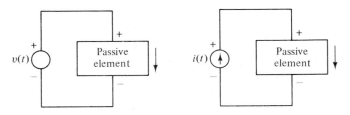

**Figure 3.16** Sources Connected to Passive Elements

into energy storage in the passive element. With this convention, if $v$ and $i$ (at any time instant) at the source terminals have the same sign, then the power supplied by the source is positive; if they are of opposite sign, then the power supplied by the source is negative (as in the charging of a battery).

### Kirchhoff's Laws

Consider a circuit consisting of a single closed loop, an example of which is shown in Figure 3.17. We know from Faraday's law (equation 2.51) that the net voltage or EMF around any closed contour is zero if there is no rate of change of flux linkage of that contour. If $i(t)$ in the network of Figure 3.17

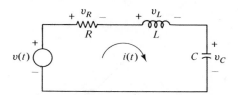

**Figure 3.17**   Single-loop Network

changes with time, and we choose a contour roughly following the wires of the circuit, there will, of course, be a changing flux linkage of this contour. In our idealization of the circuit element behavior, however, we assumed that electric fields and voltages which were caused by changing flux linkages would be considered of importance only in the inductor, not in the resistor or capacitor. For the circuit of Figure 3.17 we carry our idealization one step further, and assume that the only voltage related to changing magnetic flux linkage is the terminal voltage ($v_L$) of the inductor. The contributions to the EMF due to the electric fields in the resistor and capacitor are summarized by their terminal voltages, and the same is true for the voltage source. Assuming that no electric field exists in the connecting wires (of course, a very small field must exist if current is to flow, but the contribution of this field will be assumed negligible), Faraday's law tells us that the algebraic sum of the terminal voltages around the closed circuit must be zero. This is called *Kirchhoff's voltage law* and may be summarized by the symbolic equation

$$\sum_i v_i = 0. \tag{3.35}$$

### Example 3.3

We will use Kirchhoff's voltage law to write the equations for the circuit of Figure 3.17. In terms of the current $i(t)$, equation 3.35 yields

$$v(t) - Ri(t) - L\frac{di(t)}{dt} - \frac{1}{C}\int_{-\infty}^{t} i(\tau)\, d\tau = 0. \tag{3.36}$$

In writing this equation, we have gone around the circuit clock-wise, indicating voltage rises as positive and voltage drops as negative quantities. (This is a good rule to follow in general, but it is not essential, provided that one maintains sign consistency. We will, however, use this as the standard procedure for expressing Kirchhoff's voltage law in this text.)

Consider now a point of common interconnection among several circuit elements, as indicated in Figure 3.18. Such an interconnection point is called a *node*. (Of course, we have seen nodes before—there are four of them in the

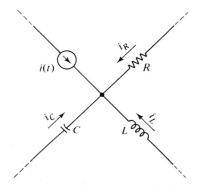

**Figure 3.18** Single Node

circuit of Figure 3.17, for example.) In our idealization of the circuit elements, we assumed that charge accumulation (i.e., capacitive phenomena) would only be associated with the capacitor, and that stray capacitance (between wires, etc.) would be neglected. With this idealization, there can be no charge accumulation at the node, except that accounted for by the capacitor. Thus the algebraic sum of the currents flowing into the node must be zero. This is called *Kirchhoff's current law* and may be summarized by the symbolic equation

$$\sum_i i_i = 0, \tag{3.37}$$

where $i_i$ are the various currents flowing either into or out of the node.

### Example 3.4

Consider the circuit of Figure 3.19. The upper node (*a*) is identical (in a topological sense) to the single node of Figure 3.18. In addition, we have another common interconnection (*b*), making this a two-node circuit. The voltage between these two nodes is indicated as $v(t)$. Writing Kirchhoff's current law, equation 3.37, we have $i(t) - i_R - i_L - i_C = 0$, or

$$i(t) - \frac{v(t)}{R} - \frac{1}{L}\int_{-\infty}^{t} v(\tau)\,d\tau - C\frac{dv(t)}{dt} = 0. \tag{3.38}$$

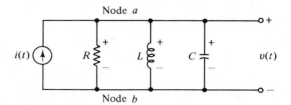

**Figure 3.19** Two-node Circuit

In this equation each term represents a current flowing *into* the upper node, or *out of* the lower node. In this text we shall adopt the standard procedure of taking the algebraic sum of currents flowing into a node and setting this sum equal to zero in order to apply Kirchhoff's current law to that node. Note that use of Kirchhoff's current law results in an equation for an unknown voltage, while use of the voltage law results in an equation for an unknown current.

### Example 3.5

Figure 3.20(a) shows a circuit containing three common interconnections or nodes, labeled *a*, *b*, and *c*. A voltage exists between each pair of these nodes, as indicated symbolically on the right side of Figure 3.20, but these

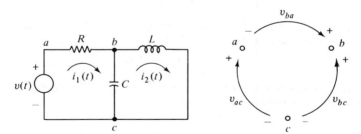

**Figure 3.20** Three-node Circuit

three voltages are not independent, because Kirchhoff's voltage law says that the sum of voltages around any closed loop must be zero. Hence

$$v_{ac} + v_{ba} - v_{bc} = 0, \tag{3.39}$$

and if any two of these voltages are known, the third may be easily determined. Let us therefore choose node *c* as a reference node, and treat $v_{ac}$ and $v_{bc}$, which are the potentials of nodes *a* and *b* with respect to *c*, as the two voltages to be found. The equation for $v_{ac}$ is trivial, since

$$v_{ac} = v(t), \tag{3.40}$$

and the source voltage $v(t)$ is specified; it is the *forcing function*. Therefore, the only unknown voltage in the circuit is $v_{bc}$; if $v_{bc}$ can be found, then the voltages across each element (the *branch voltages*) are

$$v_R = v_{ab} = v(t) - v_{bc}$$

$$v_C = v_L = v_{bc}.$$

(3.41)

We may write the equation for $v_{bc}$ by applying Kirchhoff's current law to node $b$. Setting the sum of currents flowing into node $b$ equal to zero yields

$$i_R - i_L - i_C = 0$$

$$= \frac{v(t) - v_{bc}(t)}{R} - \frac{1}{L} \int_{-\infty}^{t} v_{bc}(\tau)\, d\tau - C \frac{dv_{bc}}{dt} = 0, \qquad (3.42)$$

and the only unknown quantity in this equation is $v_{bc}$. Once $v_{bc}$ is found, the circuit is completely determined, since the branch or element voltages are known (equation 3.41), and from these the current in each element may be found from the terminal relationships.

Let us now examine this same circuit from an unknown current viewpoint. The element currents are indicated as $i_R$, $i_L$, and $i_C$ in the figure. Again, these three currents are not independent; applying Kirchhoff's current law at node $b$ yields

$$i_R - i_L - i_C = 0,$$

as we have already seen. If we choose $i_R$ and $i_L$ as the two unknown currents to be solved for, then

$$i_C = i_R - i_L.$$

This current choice is equivalent to assuming two circulating *mesh currents*, $i_1$ and $i_2$, as indicated in the figure, since then

$$i_R = i_1, \qquad i_L = i_2, \qquad i_C = i_1 - i_2.$$

As we shall see shortly, this concept of circulating mesh currents is a very convenient one and for most networks provides an independent set of currents, in terms of which each branch current may be found. If we now apply Kirchhoff's voltage law in going around the two meshes, we obtain the equations

$$v(t) - Ri_1(t) - \frac{1}{C} \int_{-\infty}^{t} [i_1(\tau) - i_2(\tau)]\, d\tau = 0$$

$$\frac{1}{C} \int_{-\infty}^{t} [i_1(\tau) - i_2(\tau)]\, d\tau - L\frac{di_2(t)}{dt} = 0.$$

(3.43)

This is a set of two simultaneous equations, in terms of the two unknown currents $i_1$ and $i_2$. If these currents may be solved for, then the network is again completely determined, since each element current and voltage may be easily determined.

Equations 3.42 and 3.43 are equally valid for determining the behavior of the circuit of Figure 3.20, and will yield identical results for the element voltages and currents. It should be noted, however, that equation 3.42 is a single equation, while equations 3.43 are a pair of simultaneous equations. Generally speaking, it is advantageous to use whichever method yields the smallest number of equations.

### Example 3.6

Figure 3.21 shows a circuit containing two independent voltage sources and a dependent current source (e.g., $R_2$ might represent the base–emitter resistance of a transistor). Treating node *d* as a reference or "datum" node, we see

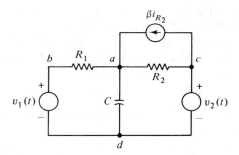

**Figure 3.21**  Four-node Circuit

that there are three other nodes which may have independent potentials with respect to this reference. Two of these *node-to-datum voltages*, however, are specified by the two voltage sources, $v_1(t)$ and $v_2(t)$. Only the potential of node *a* (with respect to datum) is unknown, and we may obtain an equation for this voltage by applying Kirchhoff's current law to node *a*. By setting the sum of the currents flowing *into* node *a* equal to zero, we obtain

$$\frac{v_1(t) - v_a(t)}{R_1} - C\frac{dv_a(t)}{dt} + \frac{v_2(t) - v_a(t)}{R_2} + \beta i_{R_2} = 0, \qquad (3.44)$$

where $v_a(t)$ is the voltage of node *a* with respect to node *d*. Since

$$i_{R_2} - \frac{v_2(t) - v_a(t)}{R_2}, \qquad (3.45)$$

equation 3.44 may be written

$$\frac{v_1(t) - v_a(t)}{R_1} - C\frac{dv_a(t)}{dt} + (1 + \beta)\frac{v_2(t) - v_a(t)}{R_2} = 0, \qquad (3.46)$$

and the only unknown quantity in this equation is $v_a(t)$.

We may also write equations for the mesh currents in this example using Kirchhoff's voltage law (Problem 3.5).

## Linear Independence of Network Voltages and Currents

Every passive element and every source in a network has associated with it a voltage (across its terminals) and a current (through the element). As we have already seen, however, all these voltages and all these currents are not linearly independent[2]; in fact, Kirchhoff's laws are simply statements of how they are *linearly dependent*. In writing equations for a network, particularly if that network is a complex one, it is desirable to choose a set of voltage or current variables which is the smallest set from which every other voltage and current in the network can be determined, in other words, the smallest complete set of linearly independent variables.

Whether or not a set of voltages or currents in a network is linearly independent depends only on the "connection configuration," or topology, of the network, and does not depend on the nature of the network elements themselves. We have already used some of the terminology associated with network topology; we will now summarize this terminology and introduce some terms:

*nodes*—interconnection points of network elements; voltages exist between pairs of nodes.

*branches*—each network element (active or passive) constitutes a branch.

*graphs*—symbolic representations of the branch-and-node topology; Figure 3.22 shows the graph of the network of Figure 3.21. Each branch of the network is represented by a line, and the nodes are the connection points of these lines. There are as many lines in the graph as there are elements in the original network.

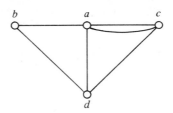

**Figure 3.22**   Graph of Network of Figure 3.21

[2]A set of variables is linearly independent if no variable of the set can be expressed as a linear combination of the other variables; i.e., if the variables $x_i$ are elements of the set $\Omega$, then the $x_i$ are linearly independent if

$$x_j \neq \sum_k a_k x_k \quad (k \neq j) \qquad \text{for all } x_j \in \Omega,$$

where $a_k$ are constants.

*trees*—graphs with the minimum number of branches removed such that no closed loops remain. Figure 3.23 shows several possible trees for the graph of Figure 3.22.

*chords*—branches that must be added to a tree to complete the graph (dashed lines in Figure 3.23).

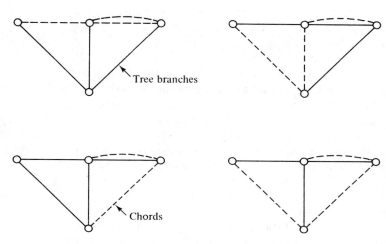

**Figure 3.23**   Some Trees of the Graph of Figure 3.22

If, in any network, we let

$$N \triangleq \text{number of nodes}$$
$$B \triangleq \text{number of branches}$$
$$T \triangleq \text{number of tree branches}$$
$$C \triangleq \text{number of chords,}$$

it is evident that

$$T = N - 1$$

and

$$C = B - T = B - (N - 1) = B - N + 1.$$

(Note that there is no fixed relationship between $N$ and $B$.)

It is not difficult to see that, if one desires to specify a minimum set of linearly independent voltages in a network, one should choose the branch voltages of branches that constitute a tree of the network. If any chord is added to a tree, a closed loop is formed, and the voltage across that chord may be determined from the tree branch voltages by using Kirchhoff's voltage law. On the other hand, if a branch is "left out" in defining the tree (i.e., if an incomplete tree is used), then one of the possible node-pair voltages is not specified. Thus the tree branch voltages always form a minimum linearly independent set of voltages. This does not, of course, imply that all these voltages will appear as unknowns in the equations—some node-pair voltages may be specified by voltage sources, as we have seen in examples. Whether "known" or "unknown," however, there will always be as many linearly independent voltages as there are tree branches.

In most networks it is convenient and customary to choose one node as a reference ("datum" or "ground") node, and to refer the potentials of all the other nodes to this reference node. In topological terms, this implies choosing a tree such that all the tree branches, if possible, "radiate" from a single node, as in the first tree of Figure 3.23. The tree branch voltages for this choice of tree are often called node-to-datum voltages, and these constitute a convenient linearly independent set of voltages to use in analyzing most networks. Each chord voltage is obviously, then, just the algebraic difference between two node-to-datum voltages.

If a network consisted only of tree branches (no chords), no current could flow in the network. If one chord is added to a tree, one closed loop is formed and one current can exist in the network. If another chord is added, two closed loops are formed, and two independent currents can flow. Following this reasoning, it becomes evident that there are as many linearly independent currents in the network as there are chords; each chord, when added to the tree, allows a new circulating loop current to flow. Again, some of these currents may be specified by current sources, but regardless of whether they are known or unknown, only $C = B - N + 1$ of them are independent. Thus the chord currents constitute a valid set of linearly independent currents to use in the network analysis.

Just as it is usually convenient to use node-to-datum voltages as a special set of tree branch voltages, it is usually convenient to choose a particular set of chord currents, called mesh currents, for the network analysis. A mesh current, however, can only be conveniently defined for a planar network, which is a network whose graph can be drawn on a plane (or, equivalently, spherical) surface without crossing lines. Figure 3.24(a) illustrates a planar graph, while the graph of Figure 3.24(b) is nonplanar. Meshes are defined, for the planar graph, as the loops that frame "window-pane-like" areas of the graph; the meshes of the graph of Figure 3.24(a) are numbered 1, 2, and 3. If we assign circulating currents to flow around these meshes, it is clear that each branch current is either simply a mesh current (if the branch belongs to only one mesh) or the algebraic difference between two mesh currents (if the branch is common to two meshes). Since we may always choose a tree such that each chord when added to that tree completes a single mesh, there are as many meshes as there are chords, and these mesh currents con-

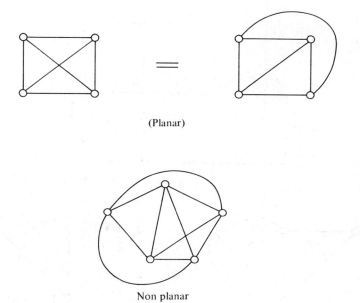

(Planar)

Non planar

**Figure 3.24**  Planar and Nonplanar Graphs

stitute a convenient set of currents for the analysis of planar networks. For nonplanar networks, the current variable choice is best made by first choosing a tree and then using the chord currents as the independent current set.

In the remainder of this text we shall consistently make use of node-to-datum voltages and (for planar networks) mesh currents in choosing variables for our network equations. This procedure, however, should be considered a special case of the more general procedure of using tree branch voltages and chord currents as variables.

***Example 3.7***

Figure 3.25 shows a network, its graph, and a node-to-datum tree. The terms $v_1$, $v_2$, and $v_3$ are the node-to-datum voltages, and $i_1$, $i_2$, $i_3$, and $i_4$ are the mesh currents. The equations for the node-to-datum voltages are

$$v_1 = v(t) \qquad \text{(specified)}$$

$$\frac{v_1 - v_2}{R_1} - \frac{1}{L_1} \int_{-\infty}^{t} v_2(\tau)\, d\tau + C \frac{d}{dt}(v_3 - v_2) + \frac{v_3 - v_2}{R_2} = 0$$

$$\frac{v_2 - v_3}{R_2} + C \frac{d}{dt}(v_2 - v_3) - \frac{1}{L_2} \int_{-\infty}^{t} v_3(\tau)\, d\tau - \frac{v_3}{R_3} = 0. \quad (3.47)$$

(The second two equations were written by setting the sums of the currents flowing *into* the nodes indicated by $v_2$ and $v_3$ equal to zero.)

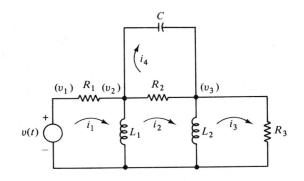

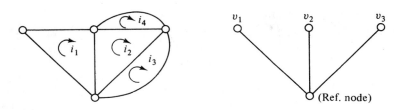

**Figure 3.25** Planar Network, Graph, and Tree

The equations for the mesh currents are

$$v(t) - R_1 i_1 - L_1 \frac{d}{dt}(i_1 - i_2) = 0$$

$$L_1 \frac{d}{dt}(i_1 - i_2) - R_2(i_2 - i_4) - L_2 \frac{d}{dt}(i_2 - i_3) = 0$$

$$L_2 \frac{d}{dt}(i_2 - i_3) - R_3 i_3 = 0$$

$$-\frac{1}{C} \int_{-\infty}^{t} i_4(\tau)\, d\tau - R_2(i_4 - i_2) = 0 \qquad (3.48)$$

(These equations were written by using Kirchhoff's voltage law, "going around" each mesh in a clockwise direction and writing voltage rises as positive and voltage drops as negative quantities.)

### Networks with Mutual Inductance

Networks containing mutual inductance are most conveniently analyzed by use of Kirchhoff's voltage law. Consider the two-mesh circuit of Figure 3.26. In going around the first mesh in a clockwise fashion, we would see a voltage drop of value $L_1(di_1/dt)$ due to the self-inductance of the coil and, in addition, a voltage induced in this coil of value $M(di_2/dt)$ due to the mutual

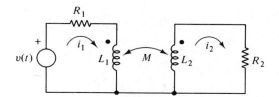

**Figure 3.26** Two-Mesh Circuit with Mutual Inductance

inductance. If the flux linkage of this coil due to $i_2$ adds to the flux linkage due to $i_1$, then a rate of change of both of these currents has the same effect; i.e., both voltage terms go into the equation with the same sign. In our example, however, the placement of the dots indicates that the flux linkages due to $i_2$ subtract from the flux linkages of $i_1$, and the effect of $di_2/dt$ is to induce a voltage in the first coil of opposite polarity to that caused by $di_1/dt$; i.e., the self- and mutual-inductance terms go into the equation with opposite signs. Thus writing the two mesh equations, we have

$$v(t) - R_1 i_1 - L_1 \frac{di_1}{dt} + M \frac{di_2}{dt} = 0$$

$$-L_2 \frac{di_2}{dt} + M \frac{di_1}{dt} - R_2 i_2 = 0. \tag{3.49}$$

(Again, we have gone clockwise around the meshes and written voltage rises as positive quantities.) Had either one of the dots or one of the current directions been reversed, the self- and mutual-inductance terms would have had the same sign in the equations, since the flux linkages due to positive $i_1$ and $i_2$ would then have added rather than subtracted.

### Example 3.8

We will write the mesh equations for the network of Figure 3.27. The different-shaped "dots" indicate the winding direction relationship for each pair of coils. Going around each mesh clockwise and using Kirchhoff's voltage law yields

$$v(t) - R_1 i_1 - L_1 \frac{di_1}{dt} + M_{12} \frac{di_2}{dt} - M_{13} \frac{di_3}{dt} = 0$$

$$-R_2 i_2 - L_2 \frac{di_2}{dt} + M_{12} \frac{di_1}{dt} - M_{23} \frac{di_3}{dt} = 0$$

$$-R_3 i_3 - L_3 \frac{di_3}{dt} - M_{13} \frac{di_1}{dt} - M_{23} \frac{di_2}{dt} = 0. \tag{3.50}$$

Note that the dot convention is equivalent to the assignment of an algebraic sign to each mutual inductance.

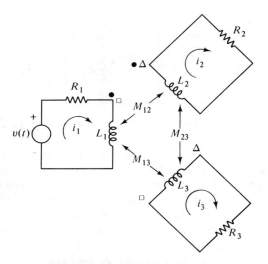

**Figure 3.27** Network with Three Coupled Coils

### Series and Parallel Element Combinations

Figure 3.28(a) shows a set of resistors connected in series. By Kirchhoff's voltage law,

$$v = R_1 i + R_2 i + \ldots + R_n i = (R_1 + R_2 + \ldots + R_n)i$$
$$= R_{eq} i, \tag{3.51}$$

where

$$R_{eq} \triangleq R_1 + R_2 + \ldots + R_n. \tag{3.52}$$

If the resistors are connected in parallel [Figure 3.28(b)], Kirchhoff's current law says that

$$i = \frac{v}{R_1} + \frac{v}{R_2} + \ldots + \frac{v}{R_n} = \left(\frac{1}{R_1} + \frac{1}{R_2} + \ldots + \frac{1}{R_n}\right)v$$
$$= \frac{1}{R_{eq}} v,$$

where

$$\frac{1}{R_{eq}} \triangleq \frac{1}{R_1} + \frac{1}{R_2} + \ldots + \frac{1}{R_n}. \tag{3.54}$$

For two resistors in parallel,

$$\frac{1}{R_{eq}} = \frac{1}{R_1} + \frac{1}{R_2} = \frac{R_1 + R_2}{R_1 R_2},$$

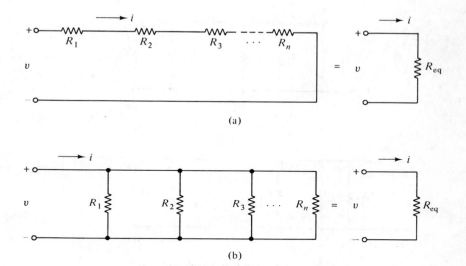

**Figure 3.28** Resistors in Series and Parallel

or

$$R_{eq} = \frac{R_1 R_2}{R_1 + R_2}.$$ (3.55)

A similar application of Kirchhoff's laws to inductors and capacitors in series and parallel gives the following formulas (Problem 3.6):

Inductors (with no mutual inductance):

Series:  $L_{eq} = L_1 + L_2 + \ldots + L_n$ (3.56)

Parallel:  $\dfrac{1}{L_{eq}} = \dfrac{1}{L_1} + \dfrac{1}{L_2} + \ldots + \dfrac{1}{L_n}.$ (3.57)

Capacitors:

Series:  $\dfrac{1}{C_{eq}} = \dfrac{1}{C_1} + \dfrac{1}{C_2} + \ldots + \dfrac{1}{C_n}$ (3.58)

Parallel:  $C_{eq} = C_1 + C_2 + \ldots + C_n.$ (3.59)

Note that resistors and inductors in series add, while capacitors in parallel add.

### Example 3.9

Consider the network of Figure 3.29(a). Parts (b) and (c) of the figure show successive simplifications by series–parallel element combinations. Why can the circuit not be simplified further?

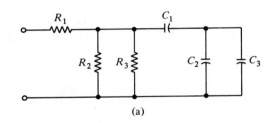

(a)

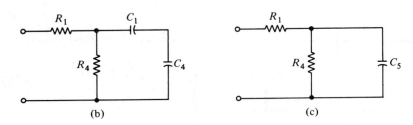

(b)          (c)

**Figure 3.29**   Network Simplification

The expressions for $R_4$, $C_4$, and $C_5$ are

$$R = \frac{R_2 R_3}{R_2 + R_3}, \qquad C_4 = C_2 + C_3, \qquad C_5 = \frac{C_1 C_4}{C_1 + C_4}.$$

In discussing elements in series and parallel, we should reiterate that voltage sources in series add, while current sources in parallel add. Ideal voltage sources in parallel or current sources in series are not allowed, for reasons previously discussed, unless their values are identical.

### Equivalent Source Representations

A physical voltage source may often be approximated by an ideal voltage source and a series resistor. Similarly, a physical current source may often be adequately approximated by an ideal current source in parallel with a resistor. By a proper choice of values, these two source representations may be made equivalent, in so far as their terminal behavior is concerned, as indicated in Figure 3.30(a). For the voltage source and series resistor, the terminal characteristic is

$$v_0 = v - R i_0. \tag{3.60}$$

For the current source and parallel resistor, the terminal characteristic is

$$i_0 = i - \frac{v_0}{R'},$$

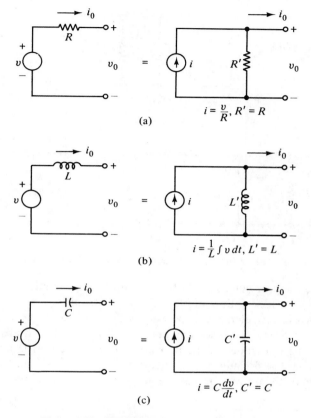

**Figure 3.30** Equivalent Source Representations

or

$$v_0 = R'i - R'i_0. \tag{3.61}$$

Thus equations 3.60 and 3.61 may be made identical, provided that $R' = R$ and $i = v/R$. Similar relationships hold for sources involving inductors and capacitors, as shown in Figure 3.30(b) and (c). (See Problem 3.7.) These source transformations may be often used advantageously to simplify networks prior to writing the network equations.

### Example 3.10

We will find the output voltage $v_0$ of the circuit of Figure 3.31(a). By doing a source transformation, we obtain the simplifications indicated in parts (b) and (c) of the figure. From part (c) it is evident that

$$v_0 = \left(i + \frac{v}{R_1}\right)\frac{R_1 R_2}{R_1 + R_2}. \tag{3.62}$$

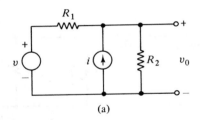

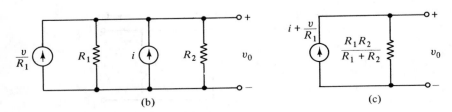

Figure 3.31 Circuit Simplification by Source Transformation

Two other source equivalences that should be kept in mind are indicated in Figure 3.32. These follow directly from the definitions of ideal sources. In any network, any element in parallel with a voltage source or in series with a current source is extraneous and can be removed with absolutely no effect upon the remainder of the network. In fact, this is true not only for single elements but for any passive element combinations (i.e., the boxes in the figure may themselves be passive networks). The extraneous elements

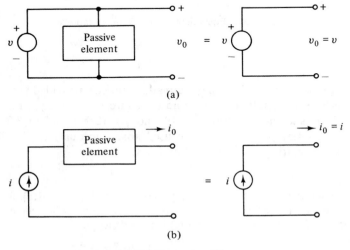

Figure 3.32 Extraneous Elements

affect only the current drawn from the ideal voltage source and the voltage across the ideal current source.

## 3.4 LINEARITY

The question of whether or not a network is linear, or whether it may at least be approximated for useful purposes by a linear mathematical model, is a very basic one. If a network cannot be treated as linear, then radically different mathematical techniques from those employed in the linear case must be used, and in general a computer simulation is required to determine the network behavior accurately. So dependent are our methods of analysis upon linear mathematics that even digital simulations of networks are usually done in terms of piecewise linear approximations. With a modern high-speed digital computer, of course, most nonlinear problems may be solved reasonably accurately. The drawbacks are that such solutions are costly, and specific digital solutions give (usually) little insight into the general nature of the solution, so that one is unable to predict behavior when parameters or configurations are changed. Since linearity is such an important concept in all engineering analysis, we will devote some discussion here to the definition of a linear network or linear system.

A resistor whose terminal characteristic is

$$v = Ri$$

is linear in the sense that if $v$ is plotted as a function of $i$, a straight line of slope $R$ results, and it is immaterial whether or not $v$ and $i$ are time functions or constants. If $i$ is considered to be the input or forcing function and $v$ the output or response, it is clear that for any linear combination of currents, $ai_1 + bi_2 + \ldots$, the response is

$$v = R(ai_1 + bi_2 + \ldots) = aRi_1 + bRi_2 + \ldots = av_1 + bv_2 + \ldots, \quad (3.63)$$

where $v_1$ is the response to $i_1$, $v_2$ the response to $i_2$, and so forth.

Consider now a capacitor, with its terminal characteristic

$$v(t) = \frac{1}{C} \int_{-\infty}^{t} i(\tau) \, d\tau. \quad (3.64)$$

Now, for any time instant $t_1$, $v(t_1)$ is not a function only of $i(t_1)$; rather, it is a function of the entire past history of $i(t)$; i.e.,

$$v(t_1) = F[i(t); (-\infty < t \leq t_1)]. \quad (3.65)$$

The capacitor (and similarly, inductor) may be said to possess a *memory*, and a network containing inductors and capacitors is often referred to as a network with memory. A plot of $v$ versus $i$ is meaningless in this case, at least if we mean a plot of instantaneous values; instantaneously, $v(t)$ is not determined by $i(t)$. If, however, we again consider a linear combination of current functions, $ai_1(t) + bi_2(t) + \dots$, our voltage response is

$$
\begin{aligned}
v(t) &= \frac{1}{C} \int_{-\infty}^{t} [ai_1(\tau) + bi_2(\tau) + \dots] \, d\tau \\
&= a \cdot \frac{1}{C} \int_{-\infty}^{t} i_1(\tau) \, d\tau + b \cdot \frac{1}{C} \int_{-\infty}^{t} i_2(\tau) \, d\tau + \dots \\
&= av_1(t) + bv_2(t) + \dots,
\end{aligned}
\tag{3.66}
$$

where $v_1(t)$ is the response to $i_1(\tau)$ for $-\infty < \tau \le t$, and so forth. Thus the linear combination of inputs results in the corresponding linear combination of outputs, and the element is called linear for this reason.

Following this argument, we may define linearity in a general way for any network or system, in terms of its forcing functions and responses. Consider a system with a single forcing-function input and a single response output, as shown in Figure 3.33. The system itself may be as simple as a single

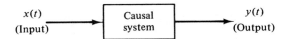

$$x(t) \qquad \boxed{\begin{array}{c} \text{Causal} \\ \text{system} \end{array}} \qquad y(t)$$
$$\text{(Input)} \qquad\qquad\qquad\qquad \text{(Output)}$$

**Figure 3.33** Single-input, Single-output System

passive element, or extremely complex, and the input and output functions may be voltages, currents, or any other variables. We will restrict our discussion to *causal systems*, meaning that the output quantity at any particular time depends only upon the present value and past history of the input, and not upon its future values (e.g., equation 3.65).

The causal system is called linear over some specified class of functions if the following input–output relations hold for all pairs of functions $x_1(\tau)$ and $x_2(\tau)$ belonging to the class:

| *Inputs* $(-\infty < \tau < t)$ | | *Outputs* (for all $t$) |
|:---:|:---:|:---:|
| $x_1(\tau)$ | $\longrightarrow$ | $y_1(t)$ |
| $x_2(\tau)$ | $\longrightarrow$ | $y_2(t)$ |
| $ax_1(\tau)$ | $\longrightarrow$ | $ay_1(t)$ |
| $x_1(\tau) + x_2(\tau)$ | $\longrightarrow$ | $y_1(t) + y_2(t)$ |

This implies the more general relationship

$$\sum_{i=1}^{N} a_i x_i(\tau) \longrightarrow \sum_{i=1}^{N} a_i y_i(t),$$

where $y_i(t)$ is the output due to the input $x_i(\tau)$. If these relationships hold, the effect of the system on the input function is said to be that of a *linear operator*. Multiplication by a constant, differentiation, and integration are all examples of elementary linear operations, and an equation that involves linear combinations of these operations is a linear integral-differential equation, provided all its terms are of first order in the variables involved. (Multiple input/multiple output systems will be discussed in a subsequent chapter.) Basically, the determination of linear network behavior and the analysis of continuous-time linear systems in general is dependent upon one's ability to find solutions to linear algebraic, differential, or integral-differential equations, and to determine how these solutions depend upon the parameters of the equations.

## PROBLEMS

**3.1** Consider a time function $f(t)$, defined by the sketch of Figure 2.26. If $f(t)$ is the current through a 1-H inductor, sketch the terminal voltage across the inductor. If $f(t)$ is the voltage across the inductor, sketch the current through the inductor. Repeat for a 1-F capacitor. Show enough information on the sketches to specify them completely.

**3.2** Discuss the analogy between the linear electrical and linear mechanical elements, when voltage is considered analogous to velocity and current analogous to force.

**3.3** A resistor, inductor, capacitor, and voltage source are all connected in series. Discuss the configuration of analogous mechanical systems, with voltage analogous to force, and with voltage analogous to velocity. What is the meaning of "mechanical node" in these two analogies?

**3.4** In the circuit of Figure 3.7, $V$ is 10 V and $I = 1$ A. Find the power supplied or dissipated by each element of the circuit if $R$ is (a) 1 $\Omega$; (b) 5 $\Omega$; (c) 10 $\Omega$.

**3.5** Write equations for the mesh currents in the circuit of Figure 3.21.

**3.6** Derive equations 3.56, 3.57, 3.58, and 3.59.

**3.7** Derive the equivalent source representations of Figure 3.30(b) and (c).

**3.8** A load resistor $R_L$ is connected across the output terminals of the small-signal common-emitter equivalent circuit of Figure 3.14. Find all the branch voltages and currents in the circuit if $v_{BE} = \sin \omega t$, $\beta = 50$, $R_R = 2000 \ \Omega$, and $R_L = 5000 \ \Omega$. Also, find the voltage and current amplifications (gains) of the circuit.

**3.9** To predict the high-frequency behavior of a common-emitter transistor circuit, the model shown in Figure 3.34 is sometimes employed.

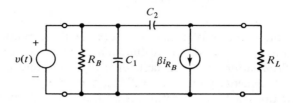

**Figure 3.34** Simplified High-frequency Transistor Model Driven by Voltage Source

Write equations for this circuit using as variables (a) mesh currents; (b) node-to-datum voltages.

**3.10** Write equations for the circuit of Figure 3.35 in terms of (a) mesh currents; (b) node-to-datum voltages.

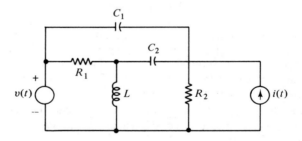

**Figure 3.35** Circuit for Problem 3.10

**3.11** Write equations for the circuit of Figure 3.36, in terms of (a) mesh currents; (b) node-to-datum voltages.

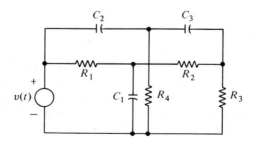

**Figure 3.36** Circuit for Problem 3.11

**3.12** Write the mesh current equations for the circuit of Figure 3.37.

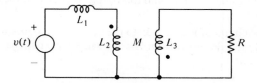

**Figure 3.37** Circuit for Problem 3.12

**3-13** Find single-element equivalent circuits for the networks shown in Figure 3.38.

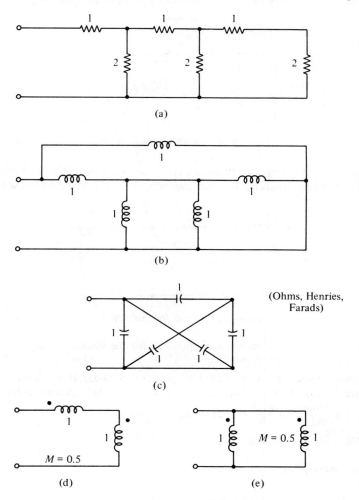

**Figure 3.38** Circuits for Problem 3.13

**3.14** If a 1-V constant-voltage source is connected across the input terminals of the circuit of Figure 3.38(a), find the voltage across, and the current through, each resistor.

**3.15** Simplify the circuit of Figure 3.39 to the point that it contains a single voltage source and a single resistor but has the same terminal characteristics as the original circuit.

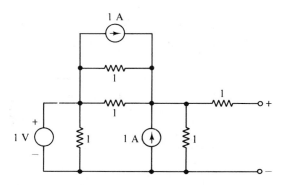

**Figure 3.39** Circuit for Problem 3.15

**3.16** Write mesh current equations for $I_1$, $I_2$, and $I_3$ in the circuit of Figure 3.40. Solve the equations for $I_3$ using determinants, and find the output voltage $v_0$.

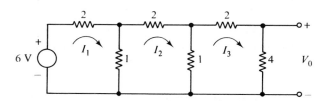

**Figure 3.40** Circuit for Problem 3.16

**3.17** Write node-to-datum voltage equations for the circuit of Figure 3.40, and solve for $v_0$ using determinants.

**3.18** State whether or not the following equations are linear. If any of the equations are nonlinear, show how they violate the basic definition of linearity:

(a) $a\dfrac{d^2 i(t)}{dt^2} + b\dfrac{di(t)}{dt} + ci(t) = v(t) + \dfrac{dv(t)}{dt}$.

(b) $\dfrac{dv(t)}{dt} + v^2(t) = 0$.

(c) $v(t)\dfrac{dv(t)}{dt} + v(t) = i(t)$.

(d) $t\dfrac{dv(t)}{dt} + v(t) = i(t) + t^2$.

(e) $a(t)\dfrac{d^2i(t)}{dt^2} = 0$.

**3.19** Write equations for the circuit of Figure 3.41. (*Hint:* Convert the current sources to equivalent voltage source representations.)

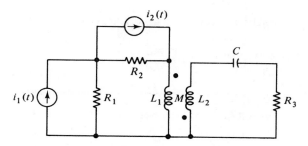

**Figure 3.41** Circuit for Problem 3.19

**3.20** In the circuit of Figure 3.42,

$$L(t) = \sin t$$
$$C(t) = \cos t.$$

Write equations for the circuit, on (a) a mesh current basis; (b) a node-to-datum voltage basis. Is the network linear?

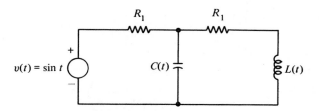

**Figure 3.42** Circuit for Problem 3.20

**3.21** The network shown in Figure 3.43 is nonplanar. Choose a set of independent current variables for this network by using chord currents, and write the network equations in terms of these currents. Identify each branch current in terms of the chosen chord currents.

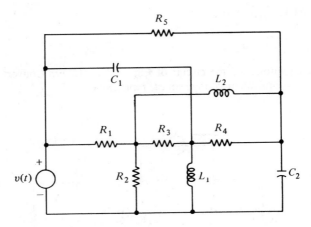

**Figure 3.43** Nonplanar Network for Problem 3.21

# 4

# *Initial Condition and*
# *Step Responses of Networks*

The behavior of networks, and systems in general, may be divided into two categories, transient behavior and steady-state behavior. These terms mean exactly what the words imply; a *transient response* is self-limited in its time duration, and after a certain period of time dies out to the point that it is no longer important. A *steady-state response* does not die out; it consists usually of time functions which are either constant in value or repeat their values in a periodic fashion. In this chapter we are concerned primarily with transient behavior, but to discuss this we must also consider networks in the constant steady state.

## 4.1 INITIAL CONDITIONS IN NETWORK ELEMENTS

### *Elements in the Constant Steady State*

In the constant steady state, nothing in a network changes with time; all voltages, currents, charges, and flux linkages are constants. Under this condition, a resistor just acts like a resistor, since Ohm's law,

$$v = Ri, \tag{4.1}$$

does not explicitly involve time, and the voltage and current are always directly proportional, whether they are constant or time-varying. For an inductor in the constant steady state, however,

$$v = L\frac{di}{dt} = 0, \tag{4.2}$$

**105**

since all derivatives must be zero, and the inductor acts exactly like a *short circuit*. The voltage across the inductor is zero, regardless of the value of the constant current flowing through it! For a capacitor, on the other hand, we have

$$i = C\frac{dv}{dt} = 0, \tag{4.3}$$

and hence the capacitor acts exactly like an *open circuit*. The current through the capacitor is zero, regardless of the value of constant voltage across its terminals! If we know, therefore, that a network is in a constant steady-state condition, we may simplify that network for purposes of analysis by replacing all the inductors by short circuits and all the capacitors by open circuits.

### Example 4.1

The circuit of Figure 4.1(a) is in the constant steady state. We wish to find the current supplied by the battery. An equivalent representation for the circuit is shown in part (b) of the figure. (Note that the mutual inductance has no effect—why?) Since there is zero voltage across $R_3$, no current flows through $R_3$, and the current supplied by the battery is just

$$I = \frac{V}{R_1 + R_2}. \tag{4.4}$$

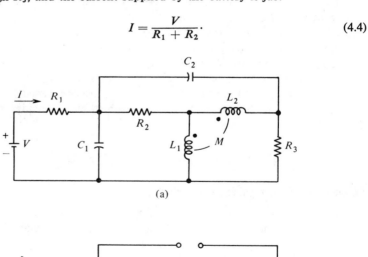

(a)

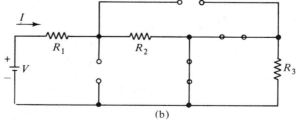

(b)

**Figure 4.1** Circuit in the Constant Steady State

### Element Response to Instantaneous Changes: Initial Conditions

In a resistor, if either the terminal voltage or current change values instantaneously, the other quantity will also, at least with our ideal-element approximation. This is not the case, however, with inductors and capacitors. If we write the inductor terminal characteristics as

$$i(t) = \frac{1}{L} \int_{-\infty}^{t} v(\tau) \, d\tau, \tag{4.5}$$

it is clear that an instantaneous jump in the voltage will not result in an instantaneous change in the current, since, by the nature of an integral, the interval over which the integral is taken must be nonzero if the value of the integral is to change, no matter how large the integrand becomes. For example, if $i(t) = v(t) = 0$ for $t < 0$, and $v(t) = V$ for $t > 0$, then $i(t) = (V/L)t$ for $t \geq 0$; the current is continuous at $t = 0$, although the voltage is not. We may summarize by stating that it is impossible to instantaneously change the current through an inductor, at least with a finite terminal voltage. Thus $i_L(0-) = i_L(0+)$. This seems very reasonable if we consider the energy stored in the inductor, which is proportional to the square of the current. To instantaneously change the current, one would have to instantaneously change the energy stored, which implies infinite power.

#### Example 4.2

A switch is in series with an inductive circuit (motor, transformer, etc.). The switch is suddenly opened. What happens?

An "ideal" switch would cause an ideal open circuit to instantaneously appear in series with the inductance. Since the inductive current cannot change to zero value instantaneously, the ideal switch is not a valid model. With a physical switch, the circuit will not be interrupted instantaneously, because an arc will form across the switch terminals. Elaborate oil circuit breakers are designed to quench this arc in large power applications where large inductive currents must be interrupted rapidly, but an instantaneous interruption of the current is not possible. Great care should always be used in opening switches in inductive circuits, since the large $di/dt$ in the inductance will cause large terminal voltages to appear.

With a capacitor, we have an analogous situation. Writing the terminal characteristic as

$$v(t) = \frac{1}{C} \int_{-\infty}^{t} i(\tau) \, d\tau, \tag{4.6}$$

it is evident, for the same reason as previously stated, that it is impossible to instantaneously change the voltage across a capacitor, at least with a finite current. Thus $v_c(0-) = v_c(0+)$. To instantaneously change the voltage, one would have to instantaneously change the charge on the capacitor, and hence its stored energy, and this would require an infinite current and power.

Let us now assume that some change takes place instantaneously in a network. This change may be a change in configuration, such as that caused by the opening or closing of switches, or a discontinuity in a source value. For convenience, we will let this change take place at the time reference $t = 0$. In order to analyze the network for $t > 0$, we shall see that it is essential to know the state of each energy storage element at the instant of time immediately following the change; i.e., at $t = 0+$. This means that we must know, or be able to find, each inductor current and each capacitor voltage (upon which the energies depend) at the time instant $t = 0+$. If we know that the network is in the constant steady state prior to $t = 0$, it is usually not difficult to evaluate the inductor currents and capacitor voltages, since we may treat the inductors as short circuits and the capacitors as open circuits. Since the current cannot change instantaneously in an inductor, for the time instant $t = 0+$ we may consider the inductor to be a current source, of value $I_L(0+) = I_L(0-)$. Similarly, since the voltage across a capacitor cannot change instantaneously, we may consider a capacitor, for $t = 0+$, to be a voltage source of value $V_c(0+) = V_c(0-)$. Inserting these appropriate sources in place of the inductors and capacitors in the network gives us an equivalent circuit valid for the time instant $t = 0+$. This equivalent circuit of course is valid only for the time instant $t = 0+$, the initial instant after the network change takes place; this is the time instant when the network is in its "initial-condition" state.

### Example 4.3

The circuit of Figure 4.2(a) is in the constant steady state prior to $t = 0$, and at $t = 0$ the switch is opened. An equivalent circuit valid for $t \leq 0$ is shown in part (b) of the figure, with the inductor replaced by a short circuit and the capacitor by an open circuit. The entire source current $I$ flows through the inductor, and the voltage across the capacitor is zero. Part (c) of the figure shows the circuit conditions for $t = 0+$. Since $i_L(0+) = i_L(0-)$ and $v_c(0+) = v_c(0-)$, the inductor is replaced by a current source of value $i_L(0-)$ as shown, and the capacitor is replaced by a voltage source of zero value (a short circuit).

### Example 4.4

The circuit of Figure 4.3(a) is in the constant steady state prior to $t = 0$, and at $t = 0$ the switch is closed. We wish to find the values of $i_1$ and $i_2$ (mesh currents) at $t = 0+$.

For $t \leq 0$, the equivalent circuit is as in part (b) of the figure. The current through $R_1$ is

$$i_{R_1}(0-) = \frac{V}{R_1 + [R_2 R_3/(R_2 + R_3)]}, \tag{4.7}$$

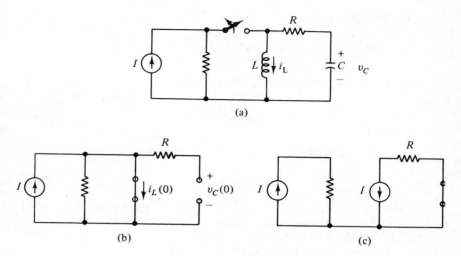

**Figure 4.2** Circuit for Example 4.3

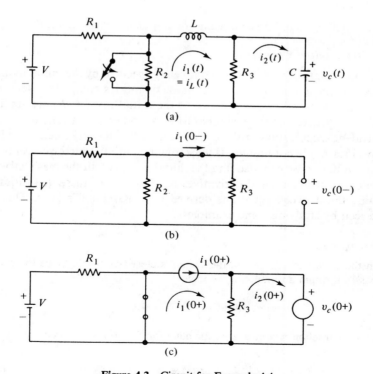

**Figure 4.3** Circuit for Example 4.4

and $v_c(0-)$ is given by

$$v_c(0-) = i_{R_1}(0-) \frac{R_2 R_3}{R_2 + R_2}. \tag{4.8}$$

$i_1(0-)$ is just the current through $R_3$, which is

$$i_1(0-) = \frac{v_c(0)}{R_3}, \tag{4.9}$$

and, since the capacitor is acting as an open circuit,

$$i_2(0-) = 0. \tag{4.10}$$

For $t = 0+$ we know that $i_L(0+) = i_1(0+) = i_1(0-)$ and that $v_c(0+) = v_c(0-)$. Thus the equivalent circuit now appears as in part (c) of the figure. Since the current (downward) through $R_3$ is just $v_c(0+)/R_3$, it follows from Kirchhoff's current law that

$$i_2(0+) = i_1(0+) - \frac{v_c(0+)}{R_3} = 0. \tag{4.11}$$

Thus both $i_1$ and $i_2$ are continuous (do not jump) when the switch is closed.

### Higher-Order Initial Conditions

The initial state of a network is completely specified by the initial inductor currents and capacitor voltages, since these quantities completely determine all the energies stored in the network at the initial instant. As we shall see shortly, however, it may be desirable to know other quantities, which give an equivalent specification of the initial state of the network; i.e., the initial-state specification is not unique. If we know the initial inductor currents and capacitor voltages, we can also find the initial values of all the mesh currents, node-to-datum voltages, and derivatives of all orders of these quantities, by examining the network equations derived from Kirchhoff's laws. This can best be seen by studying some examples.

#### Example 4.5

In the circuit of Figure 4.4, the switch is closed at $t = 0$. Since the current passes through an inductor, we have

$$i(0+) = i(0-) = 0. \tag{4.12}$$

Using Kirchhoff's voltage law, the equation for $i(t)$ is

$$L \frac{di(t)}{dt} + Ri(t) = V \qquad (t \geq 0). \tag{4.13}$$

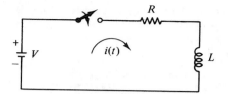

**Figure 4.4** Circuit for Example 4.5

Since this equation is valid for $t = 0+$, we have

$$L\frac{di}{dt}\bigg|_{0+} + Ri(0+) = V. \tag{4.14}$$

Thus

$$\frac{di}{dt}\bigg|_{0+} = i'(0+) = \frac{V}{L}, \tag{4.15}$$

which is the initial rate of change (or slope) of the current–time function. If we desire the initial value of the second derivative, we may obtain it by first differentiating equation 4.13; i.e.,

$$Li''(t) + Ri'(t) = 0. \tag{4.16}$$

This equation is also valid for $t = 0+$; hence

$$i''(0+) = -\frac{R}{L}i'(0+) = -\frac{R}{L}\left(\frac{V}{L}\right) = -\frac{RV}{L^2}. \tag{4.17}$$

We may continue this procedure and find the initial values of as many derivatives of the current as desired.

### Example 4.6

In the network of Figure 4.5, the initial voltage across the capacitor is specified as $v_c(0+) = +V$ volts. For $t = 0+$, the capacitor acts like a battery, giving an initial current of

$$i(0+) = -\frac{V}{R}. \tag{4.18}$$

The current equation is

$$Ri(t) + \frac{1}{C}\int_{-\infty}^{t} i(\tau)\,d\tau = 0. \tag{4.19}$$

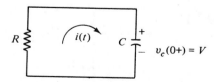

**Figure 4.5** Circuit for Example 4.6

The second term of this equation represents the voltage across the capacitor, the initial value of which is $V$. Thus equation 4.18 is seen to follow from equation 4.19, if this equation is evaluated at $t = 0+$. If we desire to find $i'(0+)$, we may differentiate equation 4.19, to obtain

$$Ri'(t) + \frac{i(t)}{C} = 0. \tag{4.20}$$

Evaluating the terms of this equation for $t = 0+$ gives

$$Ri'(0+) + \frac{i(0+)}{C} = Ri'(0+) - \frac{V}{RC} = 0. \tag{4.21}$$

Hence

$$i'(0+) = \frac{V}{R^2C}. \tag{4.22}$$

By differentiating equation 4.20, and again evaluating at $t = 0+$, we can find $i''(0+)$, and so forth.

### Example 4.7

In the network of Figure 4.6, the switch is closed at time $t = 0$, at which time there is an initial voltage $v_c(0-) = +1$ V on the capacitor. We will find the initial values (for $t = 0+$) of the two mesh currents, their first derivatives, the unknown node-to-datum voltage $v(t)$, and its derivative. Since the switch is open for $t < 0$,

$$i_1(0+) = i_L(0+) = i_L(0-) = 0. \tag{4.23}$$

Also, since

$$v_c(0+) = v(0+) = v(0-) = 1, \tag{4.24}$$

the current downward through the 1-$\Omega$ resistor is 1 A, and since $i_1(0+) = 0$, it follows that

$$i_2(0+) = -1. \tag{4.25}$$

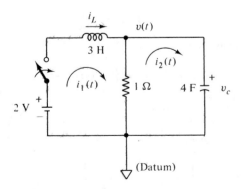

**Figure 4.6** Network for Example 4.7

The mesh current equations for $t > 0$ are

$$2 - 3\frac{di_1}{dt} - (i_1 - i_2) = 0$$

$$(i_1 - i_2) - \frac{1}{4}\int_{-\infty}^{t} i_2(\tau)\,d\tau = 0. \tag{4.26}$$

Writing these equations for the particular time instant $t = 0+$ gives

$$2 - 3i_1'(0+) - i_1(0+) + i_2(0+) = 0$$

$$i_1(0+) - i_2(0+) - v_c(0+) = 0. \tag{4.27}$$

Hence

$$2 - 3i_1'(0+) - 0 - 1 = 0,$$

or

$$i_1'(0+) = \tfrac{1}{3} \quad \text{(amperes/second)}. \tag{4.28}$$

Since $i_2'(0+)$ does not appear in the mesh equations directly, we must differentiate one of them. Differentiating the first equation of 4.26 would yield two unknowns, $i_2'(0+)$ and $i_1''(0+)$. Differentiating the second equation, however, gives (for $t = 0+$)

$$i_1'(0+) - i_2'(0+) - \tfrac{1}{4}i_2(0+) = 0, \tag{4.29}$$

and substituting known values gives

$$i_2'(0+) = \tfrac{1}{3} - \tfrac{1}{4}(-1) = \tfrac{7}{12} \quad \text{(amperes/second)}. \tag{4.30}$$

Let us now write the *node* equation for $v(t)$, using Kirchhoff's current law:

$$i_1(t) - v(t) - 4\frac{dv}{dt} = 0. \tag{4.31}$$

For $t = 0+$,

$$i_1(0+) - v(0+) = 4v'(0+),$$

or

$$v'(0+) = \tfrac{1}{4}[0 - 1] = -\tfrac{1}{4} \quad \text{(volts/second)}. \tag{4.32}$$

Thus we have found the initial values, and initial derivatives (rates of change), of $i_1$, $i_2$, and $v$. We could proceed to find as many higher-order initial conditions as desired, by successively differentiating the appropriate equations.

In the steady-state analysis of networks (such as, for example, the constant steady state or sinusoidal steady state), initial conditions play no part whatsoever. The analysis of transient behavior, however, can be thought of as consisting of three steps:

1. Writing a valid set of equations.
2. Evaluating the initial conditions.
3. Solving the equations, subject to the initial conditions and forcing functions, if any.

Generally speaking, there is no magic formula that enables one to evaluate the initial conditions in networks, without thought. A careful study of the preceding examples (and the problems at the end of the chapter), however, should clarify the general principles by which initial conditions in any lumped linear network may be evaluated.

## 4.2  SOME VERY SPECIAL TIME FUNCTIONS

### The Unit Step Function and Its Integrals

One very important type of time function in engineering analysis is the *unit step function*, which we shall designate by the symbol $u(t)$. This function has a single isolated discontinuity at the origin and may be defined as follows:

$$u(t) = \begin{cases} 0 & t < 0 \\ 1 & t > 0. \end{cases} \tag{4.33}$$

A plot of this function appears in Figure 4.7.

By using switches in conjunction with ideal voltage and current sources, we may model sources that provide step functions of voltage and current, as indicated in Figure 4.8.

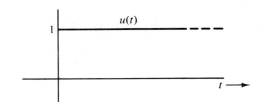

**Figure 4.7   Unit Step Function**

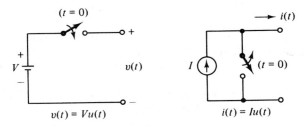

**Figure 4.8   Step-function Generators**

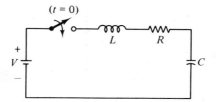

**Figure 4.9** Circuit Driven by a Voltage Step Function

Thus the equation for the circuit of Figure 4.9 may be written

$$L\frac{di(t)}{dt} + Ri(t) + \frac{1}{C} \int_{-\infty}^{t} i(\tau)\,d\tau = Vu(t), \tag{4.34}$$

$Vu(t)$ being the forcing function. (To solve the equation, it would still be necessary to know the initial value of the voltage across the capacitor. The initial value of the inductor current is zero because of the switch.)

The effect of multiplying the unit step function by a constant is obvious. In addition, we may modify the unit step by shifting it along the time axis, as illustrated by the examples of Figure 4.10. By forming linear combinations of step functions, the functions of Figure 4.11 are generated. Other time functions may be truncated in time by multiplying by step functions; some examples are shown in Figure 4.12.

Finally, other types of singularity functions may be obtained by integrating the unit step function; the first two integrals are illustrated in Figure 4.13. The first integral of the unit step is often called a *unit ramp function*,

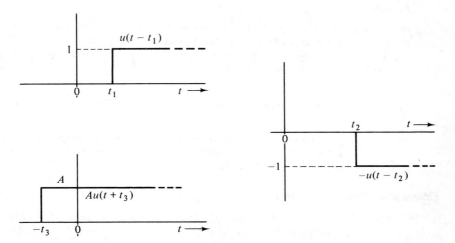

**Figure 4.10** Shifted Step Functions

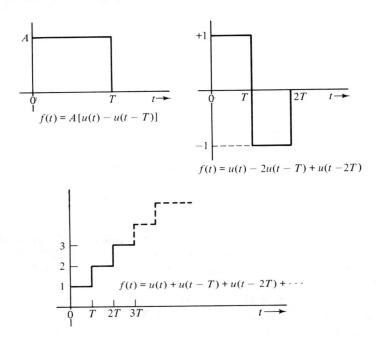

**Figure 4.11** Some Functions Formed by Linear Combinations of Step Functions

and this is also a useful function in the generation of certain waveforms of great engineering value.

### Example 4.8

In a television picture tube, the horizontal and vertical sweep circuits are required to produce deflection voltages approximating the waveform of Figure 4.14. We may write an expression for this waveform, employing the notation $u^{-1}(t)$ for the unit ramp function:

$$f(t) = \frac{A}{T}u^{-1}(t) - A[u(t - T) + u(t - 2T) + u(t - 3T) + \ldots]. \quad (4.35)$$

Note that the constant multiplying $u^{-1}(t)$ is just the slope of the ramp function. Other examples of waveforms generated using step and ramp functions appear in the problems at the end of the chapter.

### Exponential Time Functions

In lumped-linear-system analysis, one time function outranks all others in importance; this is the *exponential function* of the form $\epsilon^{st}$, where $s$ may be a real, imaginary, or complex number. If $s$ is real, say $s = \alpha$, where $\alpha$ is a real

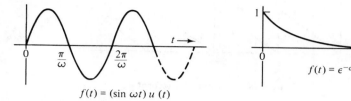

$f(t) = (\sin \omega t)\, u\,(t)$

$f(t) = \epsilon^{-\alpha t} u(t)$

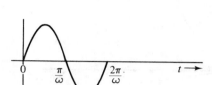

$f(t) = (\sin \omega t)\, [u(t) - u(t - \frac{2\pi}{\omega})\,]$

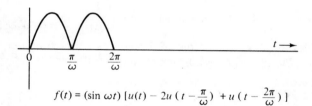

$f(t) = (\sin \omega t)\, [u(t) - 2u\,(t - \frac{\pi}{\omega}) + u\,(t - \frac{2\pi}{\omega})\,]$

**Figure 4.12**  Some Functions Formed by Multiplication by Step Functions

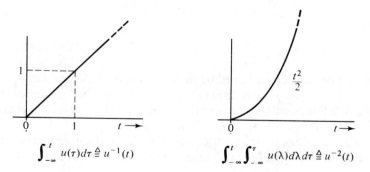

$$\int_{-\infty}^{t} u(\tau)d\tau \triangleq u^{-1}(t)$$

$$\int_{-\infty}^{t}\int_{-\infty}^{\tau} u(\lambda)d\lambda\, d\tau \triangleq u^{-2}(t)$$

**Figure 4.13**  First Two Integrals of the Unit Step Function

number, then $\epsilon^{\alpha t}$ is often referred to as a *growing* or a *decaying* exponential, depending upon whether $\alpha$ is positive or negative, respectively. A family of these functions, for various $\alpha$, is shown in Figure 4.15.

If $s$ is imaginary, say $s = j\omega$, then $\epsilon^{st}$ is a complex function; i.e., for any

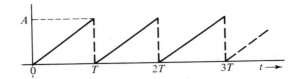

**Figure 4.14** Sawtooth Waveform

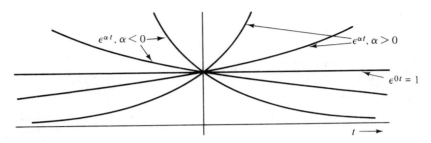

**Figure 4.15** Real Exponential Functions

value of $t$, $\epsilon^{st}$ is a complex number. The nature of this complex time function may be understood from *Euler's relationships*, which are as follows:

$$\epsilon^{j\omega t} = \cos \omega t + j \sin \omega t$$
$$\epsilon^{-j\omega t} = \cos \omega t - j \sin \omega t \tag{4.36}$$

or, conversely,

$$\cos \omega t = \frac{1}{2}(\epsilon^{j\omega t} + \epsilon^{-j\omega t})$$
$$\sin \omega t = \frac{1}{2j}(\epsilon^{j\omega t} - \epsilon^{-j\omega t}) \tag{4.37}$$

(These relationships are obtained by showing that both sides of the equations have identical Taylor's series expansions and that the Taylor's series expansion of a function is unique.)[1] From Euler's relationships, it is clear that sinusoids are special cases of exponential functions.

Certain other linear combinations of complex exponentials have special importance in solving differential equations. In particular, consider the form

$$f(t) = c\epsilon^{st} + c^*\epsilon^{s^*t}, \tag{4.38}$$

where $c$ is a complex coefficient, say $a + jb$, $s = \alpha + j\omega$, and $*$ denotes the

[1]See Appendix A.

complex conjugate. Then we have

$$f(t) = \epsilon^{\alpha t}[(a + jb)\epsilon^{j\omega t} + (a - jb)\epsilon^{-j\omega t}]$$
$$= \epsilon^{\alpha t}[(a + jb)(\cos \omega t + j \sin \omega t) + (a - jb)(\cos \omega t - j \sin \omega t)]$$
$$= \epsilon^{\alpha t}[2a \cos \omega t - 2b \sin \omega t]. \tag{4.39}$$

Note that $f(t)$ is a real function, but it only turns out to be real because the $c$ and $c^*$ coefficients are a conjugate pair, so all the imaginary terms in equation 4.39 cancel. Since, by trigonometric identities,

$$a \cos \omega t - b \sin \omega t = \sqrt{a^2 + b^2} \cos\left(\omega t + \tan^{-1} \frac{b}{a}\right), \tag{4.40}$$

we may also write equation 4.39 as

$$f(t) = 2\sqrt{a^2 + b^2}\, \epsilon^{\alpha t} \cos\left(\omega t + \tan^{-1} \frac{b}{a}\right). \tag{4.41}$$

Equation 4.41 indicates a cosine waveform, modified in two ways:

1. It is shifted in the negative direction by the phase angle $\tan^{-1}(b/a)$.
2. It is multiplied by either a growing or decaying exponential, depending on the sign of $\alpha$. [Of course, if $\alpha = 0$, $f(t)$ is a pure sinusoid; i.e,. it neither grows nor decays with time.]

A sketch of $f(t)$, for $\alpha < 0$, is indicated in Figure 4.16. The importance of the role played by the exponential functions in the analysis of lumped linear networks will become evident in the next section.

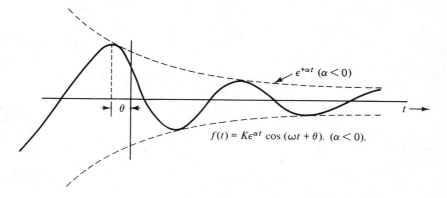

**Figure 4.16**  Exponentially Damped Sinusoid

## 4.3 TRANSIENT RESPONSE OF FIRST-ORDER NETWORKS

### *First-Order Networks: Initial Condition Response*

First-order networks are defined to be those whose behavior is described by a single (scalar) first-order differential equation. In this section we shall investigate the possible responses of such a network in the absence of any forcing or driving functions, i.e., the responses to initial conditions only.

The simplest examples of first-order circuits are shown in Figure 4.17.

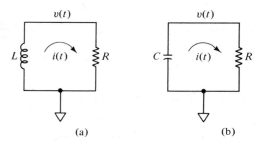

**Figure 4.17** Simplest First-order Networks

Writing the equations for the mesh currents for the circuits of (a) and (b), respectively, we have

$$\text{(a)} \quad L\frac{di}{dt} + Ri = 0. \tag{4.42}$$

$$\text{(b)} \quad Ri + \frac{1}{C}\int_{-\infty}^{t} i(\tau)\, d\tau = 0, \tag{4.43}$$

or differentiating,

$$\text{(b)} \quad R\frac{di}{dt} + \frac{1}{C}i = 0. \tag{4.44}$$

Similarly, writing equations for the node voltages $v(t)$ gives

$$\text{(a)} \quad \frac{1}{L}\int_{-\infty}^{t} v(\tau)\, d\tau + \frac{1}{R}v = 0, \tag{4.45}$$

or differentiating,

$$\frac{1}{R}\frac{dv}{dt} + \frac{1}{L}v = 0 \tag{4.46}$$

$$\text{(b)} \quad C\frac{dv}{dt} + \frac{1}{R}v = 0. \tag{4.47}$$

Each of these first-order differential equations is of the general form

$$\frac{dx}{dt} + \alpha x = 0, \tag{4.48}$$

where $\alpha$ is either $R/L$ or $1/RC$, and this equation may be solved directly by separation of variables. Rearranging terms,

$$\frac{dx}{x} = -\alpha \, dt. \tag{4.49}$$

Integrating both sides gives

$$\int \frac{dx}{x} = \log_\epsilon x + K_1 = -\alpha \int dt = -\alpha t + K_2. \tag{4.50}$$

Combining together the two constants of integration, we may write

$$\log_\epsilon x = K_3 - \alpha t, \tag{4.51}$$

or

$$x(t) = K\epsilon^{-\alpha t}, \tag{4.52}$$

where $K_3 = \log_\epsilon K$. Equation 4.52 represents an infinite family of possible solutions, depending upon the value of $K$. (Note, however, that for the circuits shown, with $R$, $L$, and $C$ positive, the solutions are all decaying exponentials.) If, for any $t = t_1$, $x(t_1)$ is specified, then $K$ is determined, and this specifies one particular solution out of the infinite family. In particular, for $t = 0$, $x(0) = K$, so $K$ is simply the initial condition at $t = 0$.

### Example 4.9

In the circuit of Figure 4.18, the switch is thrown in the direction of the arrow at $t = 0$. If the circuit was in the constant steady state prior to $t = 0$, then

$$i(0-) = i(0+) = \frac{V}{R}. \tag{4.53}$$

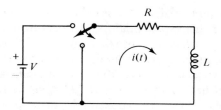

**Figure 4.18**  Circuit for Example 4.9

This is the required initial condition; for $t > 0$,

$$i(t) = \frac{V}{R} \epsilon^{-(R/L)t} \qquad (t > 0). \tag{4.54}$$

### Example 4.10

In the circuit of Figure 4.19, the switch is opened at $t = 0$, the circuit previously being in the constant steady state. The initial conditions on $v_c(t)$ and $i(t)$ are

$$v_c(0-) = v_c(0+) = V \tag{4.55}$$

$$i(0+) = -i_R(0+) = -\frac{V}{R}. \tag{4.56}$$

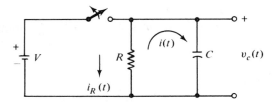

**Figure 4.19**  Circuit for Example 4.10

Therefore, for $t > 0$, the solutions for $v_c(t)$ and $i(t)$ are

$$v_c(t) = V\epsilon^{-t/RC} \qquad (t > 0) \tag{4.57}$$

$$i(t) = -\frac{V}{R}\epsilon^{-t/RC} \qquad (t > 0). \tag{4.58}$$

### Time Constant

It is common practice to write a real exponential, like that of equation 4.52, in terms of the *time constant*, $\tau$, defined by

$$\tau = \frac{1}{\alpha}. \tag{4.59}$$

If we examine the exponential, $\epsilon^{-t/\tau}$, it is evident that after an interval of one time constant, $t = \tau$, the exponential has decayed from a value of unity at $t = 0$ to the value $\epsilon^{-1}$, which is approximately 0.3679. Thus the exponential, after one time constant, has gone about 63% of the way from its initial value at $t = 0$ to its final value as $t \to \infty$. After five time constants, $\epsilon^{-5} = 0.0067$, and the exponential has completed about 99.3% of its total excursion—in most practical applications, the exponential can be considered over after five or six time constants.

Since, for a first-order network, $\alpha$ is either $R/L$ or $1/RC$, it follows that for:

$$RL \text{ networks:} \quad \tau = \frac{L}{R}$$

$$RC \text{ networks:} \quad \tau = RC.$$

Dimensionally, both $L/R$ and $RC$ have the dimensions of time.

### First-Order Networks: Forced Response

Let us now return to the first-order differential equation, equation 4.48. This equation, with the right-hand side equal to zero, is called a *homogeneous equation*, since the only time function involved in the equation is $x(t)$. Now, however, we wish to include a forcing function on the right-hand side of the equation, as in equation 4.60:

$$\frac{dx}{dt} + \alpha x = f(t). \tag{4.60}$$

This equation may be integrated if we first multiply both sides by $\epsilon^{\alpha t}$, which is called an *integrating factor*, to give

$$\epsilon^{\alpha t} \frac{dx}{dt} + \alpha x \epsilon^{\alpha t} = \epsilon^{\alpha t} f(t). \tag{4.61}$$

From the rule for differentiating a product, it is clear that

$$\frac{d}{dt}(x\epsilon^{\alpha t}) = \epsilon^{\alpha t} \frac{dx}{dt} + \alpha x \epsilon^{\alpha t}. \tag{4.62}$$

Therefore,

$$\frac{d}{dt}(x\epsilon^{\alpha t}) = \epsilon^{\alpha t} f(t), \tag{4.63}$$

and this equation may be integrated directly to give

$$x\epsilon^{\alpha t} = \int_{t_1}^{t} \epsilon^{\alpha \tau} f(\tau)\, d\tau + K_1, \tag{6.64}$$

where $K_1$ is a constant of integration. Since normally we are interested in forcing functions that are applied at $t = 0$, we will write the lower limit of the integral as zero; i.e.,

$$x\epsilon^{\alpha t} = \int_{0}^{t} \epsilon^{\alpha \tau} f(\tau)\, d\tau + K. \tag{4.65}$$

Multiplying both sides of this equation by $\epsilon^{-\alpha t}$ gives

$$x(t) = K\epsilon^{-\alpha t} + \int_0^t \epsilon^{-\alpha(t-\tau)} f(\tau) \, d\tau. \tag{4.66}$$

Note that the first term of this solution is identical to the solution of the homogeneous equation and involves the constant $K$, which is determined by the initial conditions. The second term, called a *convolution integral*, gives the response due to the forcing function $f(t)$ and does not involve the initial conditions. Convolution integrals will be discussed in more detail in subsequent chapters.

### Example 4.11

In the circuit of Figure 4.20,

$$v(t) = \epsilon^{-\beta t} u(t). \tag{4.67}$$

Find $i(t)$.

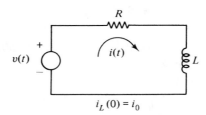

$$i_L(0) = i_0$$

**Figure 4.20**   Circuit for Example 4.11

Writing the mesh equation, we have

$$L\frac{di}{dt} + Ri = v(t), \tag{4.68}$$

or

$$\frac{di}{dt} + \frac{R}{L}i = \frac{1}{L}\epsilon^{-\beta t} u(t). \tag{4.69}$$

Since $v(t) = 0$ for $t < 0$, we may use equation 4.66 directly:

$$i(t) = K\epsilon^{-(R/L)t} + \int_0^t \frac{1}{L}\epsilon^{-(R/L)(t-\tau)}\epsilon^{-\beta\tau} u(\tau) \, d\tau$$

$$= K\epsilon^{-(R/L)t} + \frac{1}{L}\epsilon^{-(R/L)t} \int_0^t \epsilon^{[(R/L)-\beta]\tau} \, d\tau$$

$$= K\epsilon^{-(R/L)t} + \frac{1}{[(R/L)+\beta]L}\epsilon^{-(R/L)t}(\epsilon^{[(R/L)-\beta]t} - 1)$$

$$= K\epsilon^{-(R/L)t} + \frac{1}{R - \beta L}[\epsilon^{-\beta t} - \epsilon^{-(R/L)t}] \qquad (t > 0). \tag{4.70}$$

Since $i(t) = i_0$ at $t = 0$, $K = i_0$, and

$$i(t) = i_0 \epsilon^{-(R/L)t} + \frac{1}{R - \beta L} [\epsilon^{-\beta t} - \epsilon^{-(R/L)t}] \qquad (t > 0). \qquad (4.71)$$

### Example 4.12

In the circuit of Figure 4.21,

$$i(t) = (\sin t)u(t). \qquad (4.72)$$

Find $v(t)$.

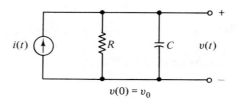

**Figure 4.21** Circuit for Example 4.12

Writing the node equation, we have

$$C\frac{dv}{dt} + \frac{v}{R} = i(t), \qquad (4.73)$$

or

$$\frac{dv}{dt} + \frac{1}{RC}v = \frac{1}{C}\sin t \, u(t). \qquad (4.74)$$

Using equation 4.66,

$$v(t) = K\epsilon^{-(t/RC)} + \int_0^t \frac{1}{C} \epsilon^{-(1/RC)(t-\tau)} \sin \tau \, u(\tau) \, d\tau$$

$$= K\epsilon^{-(t/RC)} + \frac{1}{C}\epsilon^{-(t/RC)} \int_0^t \epsilon^{(\tau/RC)} \sin \tau \, d\tau$$

$$= K\epsilon^{-t/RC} + \frac{1}{C} \epsilon^{-t/RC} \frac{R^2 C^2}{1 + R^2 C^2} \left[ \epsilon^{t/RC} \left( \frac{1}{RC}\sin t - \cos t \right) + 1 \right]$$

$$(t > 0). \qquad (4.75)$$

Since $v(0) = v_0$, $K = v_0$, and

$$v(t) = v_0 \epsilon^{-t/RC} + \frac{R^2 C}{1 + R^2 C^2} \left( \frac{1}{RC}\sin t - \cos t + \epsilon^{-t/RC} \right) \qquad (t > 0). \qquad (4.76)$$

## First-Order Networks: Step Response

In this section we shall concentrate on one particular first-order forced response—the response of a first-order network to a step-function input.

In this case $f(t) = K_1 u(t)$, and equation 4.66 becomes

$$x(t) = K\epsilon^{-\alpha t} + K_1 \int_0^t \epsilon^{-\alpha(t-\tau)} u(\tau)\, d\tau$$

$$= K\epsilon^{-\alpha t} + K_1 \epsilon^{-\alpha t} \int_0^t \epsilon^{\alpha\tau}\, d\tau$$

$$= K\epsilon^{-\alpha t} + \frac{K_1}{\alpha} \epsilon^{-\alpha t}(\epsilon^{\alpha t} - 1)$$

$$= K\epsilon^{-\alpha t} + \frac{K_1}{\alpha}(1 - \epsilon^{-\alpha t}) \qquad (t > 0), \qquad (4.77)$$

with $K = x(0)$. The importance of this particular forcing function is that it represents the sudden application of a constant voltage or current source to a network, as indicated in the following examples.

### Example 4.13

In the circuit of Figure 4.22, the switch is closed at $t = 0$. What is $i(t)$?
The mesh equation may be written as

$$L\frac{di}{dt} + Ri = Vu(t), \qquad (4.78)$$

or

$$\frac{di}{dt} + \frac{R}{L}i = \frac{V}{L}u(t). \qquad (4.79)$$

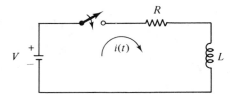

**Figure 4.22**   Circuit for Example 4.13

Applying equation 4.77, we have

$$i(t) = K\epsilon^{-(R/L)t} + \frac{V}{R}[1 - \epsilon^{-(R/L)t}] \qquad (t > 0). \qquad (4.80)$$

Since $i(0+) = i(0-) = 0$, $K = 0$, and hence

$$i(t) = \frac{V}{R}[1 - \epsilon^{-(R/L)t}] \qquad (t > 0). \qquad (4.81)$$

A sketch of this current is shown in Figure 4.23.

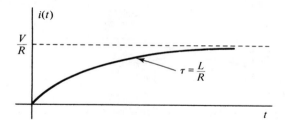

**Figure 4.23**  Current in Circuit of Figure 4.22

## Example 4.14

In the circuit of Figure 4.24(a), the switch is closed at $t = 0$, the circuit previously having been in the constant steady state. What is $i(t)$?

An equivalent representation for this circuit is indicated in part (b) of the figure, with the initial inductor current being

$$i(0+) = i_L(0+) = i_L(0-) = i_0 = \frac{V}{R_1 + R_2}. \tag{4.82}$$

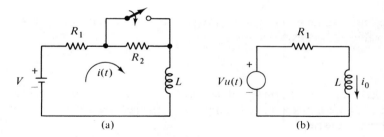

**Figure 4.24**  Circuit for Example 4.14

The equation valid for $t > 0$ is

$$\frac{di}{dt} + \frac{R_1}{L}i = \frac{V}{L}u(t), \qquad i_0 = \frac{V}{R_1 + R_2}. \tag{4.83}$$

From equation 4.77,

$$i(t) = \frac{V}{R_1 + R_2}\epsilon^{-(R_1/L)t} + \frac{V}{R_1}[1 - \epsilon^{-(R_1/L)t}] \qquad (t > 0). \tag{4.84}$$

Thus the current has the value $V/(R_1 + R_2)$ at $t = 0+$ and approaches the value $V/R_1$ as $t \to \infty$ (as it should, since in both cases the inductor behaves like a short circuit) and is an exponential of time constant $L/R_1$; this function is sketched in Figure 4.25.

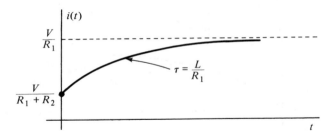

**Figure 4.25**  Current Response in Example 4.14

In networks that involve simple first-order transients, it is not necessary to write and solve the equation each time in order to sketch and write the mathematical form of the solution. All that is required, in fact, is a knowledge of the initial condition (at $t = 0+$), the final condition (as $t \to \infty$), and the time constant. These three pieces of information completely specify the exponential. If the exponential has an initial value $x_0$ and decays away to zero with a time constant $\tau$, the expression for that exponential is

$$x(t) = x_0 \epsilon^{-t/\tau}. \tag{4.85}$$

If the initial value is zero and the exponential builds up toward a constant final value $x_\infty$, the expression is

$$x(t) = x_\infty(1 - \epsilon^{-t/\tau}) \tag{4.86}$$

as in Example 4.13. If, on the other hand, the initial value is $x_0$ and the final value $x_\infty$, both nonzero as in Example 4.14, the general expression (equation 4.77) may be written

$$x(t) = x_0 \epsilon^{-t/\tau} + x_\infty(1 - \epsilon^{-t/\tau}). \tag{4.87}$$

Equation 4.87 thus represents the most general form of response to an instantaneous change in a simple first-order network.

### Example 4.15

In the circuit of Figure 4.26(a), the switch is thrown in the direction of the arrow at $t = 0$, the circuit previously having been in the constant steady state. What is $v(t)$?

The three pertinent pieces of information in determining the solution are

$$v(0+) = v(0-) = 0$$

$$v(\infty) = IR \quad \text{(the capacitor acting like an open circuit)}$$

and

$$\tau = RC.$$

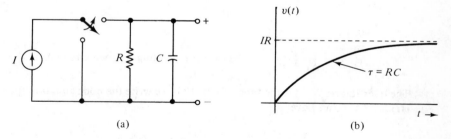

**Figure 4.26**  Circuit for Example 4.15

Hence

$$v(t) = IR(1 - \epsilon^{-t/\tau}) \qquad (t > 0), \tag{4.88}$$

as indicated in part (b) of the figure.

### First-Order Networks with More Than Two Elements

A network may contain more than two passive elements and yet still be of first order, as in the following example:

#### Example 4.16

In the circuit of Figure 4.27(a), the switch is closed at $t = 0$, with the circuit previously being in the constant steady state. What is $v(t)$? The initial and final values of $v(t)$ are easily found from the constant-steady-state equivalent

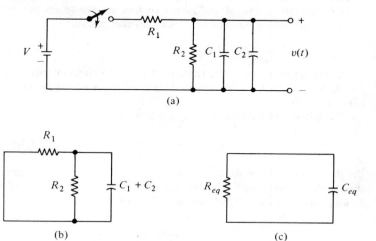

**Figure 4.27**  Circuit and Its Equivalents for Time-constant Evaluation

circuits:

$$v(0+) = v(0-) = 0$$

$$v(\infty) = \frac{VR_2}{R_1 + R_2} \qquad \text{(the capacitors acting as open circuits).}$$

The question is: What is the time constant? If we write the node equation for $v(t)$ for $t > 0$, we have

$$\frac{v - V}{R_1} + \frac{v}{R_2} + C_{eq}\frac{dv}{dt} = 0 \qquad (t > 0) \tag{4.89}$$

$$C_{eq} \triangleq C_1 + C_2.$$

Thus

$$\frac{v}{R_1} + \frac{v}{R_2} + C_{eq}\frac{dv}{dt} = \frac{V}{R_1} \qquad (t > 0). \tag{4.90}$$

This may be written as

$$\frac{dv}{dt} + \frac{1}{R_{eq}C_{eq}}v = \frac{V}{R_1 C_{eq}} \qquad (t > 0), \tag{4.91}$$

where

$$R_{eq} \triangleq \frac{R_1 R_2}{R_1 + R_2}.$$

Thus, for evaluating the time constant, we should replace the voltage source by a short circuit, which effectively puts the two resistors in parallel, as indicated in parts (b) and (c) of the figure. (An ideal voltage source has zero internal resistance, hence the replacement by a short circuit.)

### Example 4.17

In the circuit of Figure 4.28, the circuit is in the constant steady state for $t < 0$, and at $t = 0$ the switch is thrown in the arrow direction. What is $i_L(t)$ for $t > 0$?

The initial and final values of $i_L$ are easily found:

$$i_L(0+) = i_L(0-) = 0;$$

the final value may be found by replacing the inductor by a short circuit—the current $I$ then simply divides, part going through $R_1$ and the remainder through $R_2$ (and $L$), inversely as the resistance values. Thus

$$i_L(\infty) = i_{R_2}(\infty) = I\frac{R_1}{R_1 + R_2}.$$

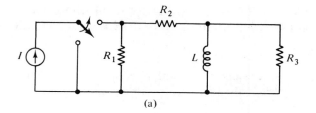

(a)

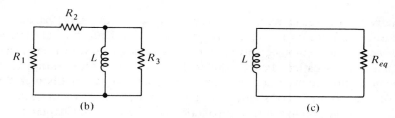

(b)                                      (c)

**Figure 4.28**   Circuit and Its Equivalents for Time-constant
Evaluation

To evaluate the time constant, we should replace the current source by an
open circuit (since an ideal current source has infinite internal resistance—
i.e., the current does not change regardless of the terminal voltage). Hence
the reduction proceeds as in parts (b) and (c) of the figure, and

$$\tau = \frac{L}{R_{eq}},$$

where

$$R_{eq} = \frac{(R_1 + R_2)R_3}{R_1 + R_2 + R_3}.$$

Thus $i_L(t)$ is

$$i_L(t) = \frac{IR_1}{R_1 + R_2}(1 - \epsilon^{-t/\tau}). \tag{4.92}$$

In summary, not all *RC* or *RL* circuits are first order; in general, they will
only be of first order if all the like elements in the circuit can be combined
into single-element equivalents, when voltage sources are replaced by short
circuits and current sources by open circuits, which procedure allows one to
evaluate the time constant. Examples of *RC* and *RL* circuits that are not of
first order will be found in the problems at the end of the chapter.

## 4.4  INITIAL-CONDITION RESPONSE OF
## SECOND- AND HIGHER-ORDER NETWORKS

### *Homogeneous Equations*

We shall now consider the solution of an $n^{\text{th}}$-order linear, ordinary, homogeneous differential equation, i.e., an equation of the form

$$a_n \frac{d^n x}{dt^n} + a_{n-1} \frac{d^{n-1} x}{dt^{n-1}} + \ldots + a_0 x(t) = 0. \tag{4.93}$$

We have already seen a number of examples of such equations, particularly ones of second order ($n = 2$), in previous examples. Unfortunately, higher-order equations cannot be solved in quite as straightforward a manner as the first-order equation. In fact, this is the reason for concentrating at this time on homogeneous equations (no forcing functions)—in Chapter 5 we discuss one type of forced solution of special importance, but a more general treatment of the nonhomogeneous equation is deferred until Chapter 7.

It may be shown[2] that for any finite set of initial conditions, the solution of an equation of the form of (4.93) is unique. This means that if one can find *a* solution, by any possible means, it is also *the* solution—two different solutions cannot result from the same set of initial conditions. If we examine equation 4.93, we note that a linear combination of $x(t)$ and its derivatives must be zero for all values of $t$; this will only be possible if $x(t)$ and its derivatives have the same general functional form. (For example, two terms, one involving a real exponential and the other a sinusoid, might add to zero for certain values of $t$ but clearly cannot add to zero for *all* values of $t$.) Thus we are led to look for a solution such that the function itself and its derivatives of all orders have the same functional form. There is only one general type of function which has the property that it may be differentiated (or integrated) any number of times and still retain the same functional form. This is the *complex exponential function* (of which real exponentials and sinusoids are special cases), which we have discussed previously. Thus we are led to the complex exponential as a trial solution for equation 4.93, bearing in mind that if we can find one solution we need look no further, since it is unique.

If we substitute the trial solution

$$x(t) = \epsilon^{st} \tag{4.94}$$

into equation 4.93, we obtain

$$a_n s^n \epsilon^{st} + a_{n-1} s^{n-1} \epsilon^{st} + \ldots + a_0 \epsilon^{st} = 0. \tag{4.95}$$

[2]See, for example, E. A. Coddington and N. Levinson, *Theory of Ordinary Differential Equations*, McGraw-Hill Book Company, New York, 1955.

Dividing by $\epsilon^{st}$, we obtain an equation in the variable $s$:

$$a_n s^n + a_{n-1} s^{n-1} + \ldots + a_0 = 0. \qquad (4.96)$$

This is an algebraic equation of $n^{\text{th}}$ order in $s$ and is called the *characteristic equation*. By the fundamental theorem of algebra, this equation must have exactly $n$ roots, $s_1, s_2, \ldots, s_n$, and these determine the allowable set of exponentials, or the exponentials that will satisfy equation 4.93. Since any linear combination of these exponentials will also satisfy equation 4.93 (why?), we may write the most general solution as

$$x(t) = K_1 \epsilon^{s_1 t} + K_2 \epsilon^{s_2 t} + \ldots + K_n \epsilon^{s_n t}, \qquad (4.97)$$

where the $K$'s are constants that must be determined from the initial conditions.

### Example 4.18

In the circuit of Figure 4.29, the switch is closed at $t = 0$, with no initial charge on the capacitor. What is $i(t)$?

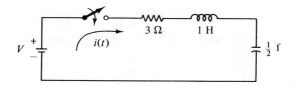

**Figure 4.29**  Circuit for Example 4.18

With the switch closed, the circuit mesh equation is

$$\frac{di}{dt} + 3i + 2 \int_{-\infty}^{t} i(\tau)\, d\tau = V. \qquad (4.98)$$

Because of the inductor,

$$i(0+) = i(0-) = 0, \qquad (4.99)$$

and since there is no initial voltage on the capacitor, the initial value of the integral in equation 4.98 is zero, so

$$\left. \frac{di}{dt} \right|_{0+} = V. \qquad (4.100)$$

If we now differentiate equation 4.98 once, we get

$$\frac{d^2 i}{dt^2} + 3\frac{di}{dt} + 2i(t) = 0. \qquad (4.101)$$

This is a second-order homogeneous equation (note that the constant forcing function on the right-hand side of equation 4.98 is eliminated in the process of differentiating). The characteristic equation is

$$s^2 + 3s + 2 = 0, \tag{4.102}$$

and its roots are $s_1 = -1$ and $s_2 = -2$. The general form of the solution is therefore

$$i(t) = K_1\epsilon^{-t} + K_2\epsilon^{-2t}, \tag{4.103}$$

and $K_1$ and $K_2$ are obtained from the initial conditions as follows:

$$i(0+) = K_1 + K_2 = 0$$

$$\left.\frac{di}{dt}\right|_{0+} = -K_1\epsilon^{-t} - 2K_2^{-2t}\Big|_{0+} = -K_1 - 2K_2 = V. \tag{4.104}$$

Solving this pair of equations for $K_1$ and $K_2$ gives

$$K_1 = -K_2 = V. \tag{4.105}$$

The solution for $i(t)$ is therefore

$$i(t) = V(\epsilon^{-t} - \epsilon^{-2t}). \tag{4.106}$$

The general form of this solution is sketched in Figure 4.30.

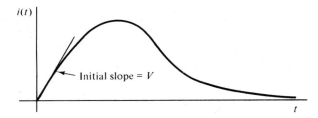

**Figure 4.30**   Sketch of Solution of Example 4.18

*Example 4.19*

In the circuit of Figure 4.31, the initial currents are given as follows:

$$i_1(0+) = I_1, \qquad i_2(0+) = I_2.$$

Find $i_1(t)$ for $t > 0$.

The mesh equations for the circuit may be written

$$i_1 + \frac{d}{dt}(i_1 - i_2) = 0$$

$$\frac{d}{dt}(i_1 - i_2) - i_2 - \frac{di_2}{dt} = 0. \tag{4.107}$$

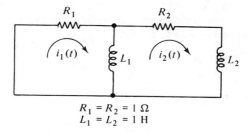

$$R_1 = R_2 = 1 \ \Omega$$
$$L_1 = L_2 = 1 \ H$$

**Figure 4.31** Circuit for Example 4.19

To find $i_{L_1}(t)$, we need to reduce equations 4.107 to a single equation in $i_1$. The easiest way to do this is to introduce the operator $D$ for $d/dt$, so that

$$Dx \triangleq \frac{dx}{dt}$$

$$D^2x = \frac{d^2x}{dt^2}$$

$$\frac{1}{D}x = \triangleq \int x \ dt, \qquad \text{etc.}$$

This device enables us to keep track of the various orders of derivatives and integrals in reducing a set of differential-integral equations to a single equation in a desired unknown, since the $D$ operator may be manipulated algebraically. Thus we may write equations 4.107 as

$$(D + 1)i_1 - Di_2 = 0$$
$$Di_1 - (2D + 1)i_2 = 0. \tag{4.108}$$

Solving for $i_1$ gives

$$D^2i_1 + 3Di_1 + i_1 = 0, \tag{4.109}$$

or

$$\frac{d^2i_1}{dt^2} + 3\frac{di_1}{dt} + i_1 = 0. \tag{4.110}$$

The characteristic equation associated with (4.110) is

$$s^2 + 3s + 1 = 0, \tag{4.111}$$

which has the roots (to three significant places)

$$s_1 = -0.38, \qquad s_2 = -2.62.$$

Thus the general solution is

$$i_1(t) = K_1\epsilon^{-0.38t} + K_2\epsilon^{-2.62t}. \tag{4.112}$$

To find $K_1$ and $K_2$ we need to know $i_1(0+)$ and $di_1/dt \mid_{0+}$. $i_1(0+)$ is given as $I_1$. To find $di_1/dt \mid_{0+}$ we may go back to the original mesh equations (4.107). From the first equation, evaluated at $t = 0+$, we have

$$\frac{d}{dt}(i_1 - i_2)\bigg|_{0+} = -i_1(0+) = -I_1.$$

From the second equation,

$$\frac{di_2}{dt}\bigg|_{0+} = \frac{d}{dt}(i_1 - i_2)\bigg|_{0+} - i_2(0+) = -I_1 - I_2.$$

Finally, from the first equation again,

$$\frac{di_1}{dt}\bigg|_{0+} = \frac{di_2}{dt}\bigg|_{0+} - i_1(0+) = -I_1 - I_2 - I_1 = -2I_1 - I_2.$$

Putting in these initial conditions, we have

$$\begin{aligned} i_1(0+) &= K_1 + K_2 = I_1, \\ \frac{di_1}{dt}\bigg|_{0+} &= -0.38K_1 - 2.62K_2 = -2I_1 - I_2, \end{aligned} \tag{4.113}$$

and these two equations may be easily solved for $K_1$ and $K_2$.

Note that the network in this example is a second-order one, even though it contains only resistors and inductors. [A second-order equation may also be obtained, in similar fashion, for $i_2(t)$.]

### Example 4.20

Repeat the problem of Example 4.18, with the element values in the circuit of Figure 4.29 changed as follows: $L = 1$ H, $R = 2\ \Omega$, $C = \frac{1}{2}$ F. The mesh equation is then

$$\frac{di}{dt} + 2i + 2\int_{-\infty}^{t} i(\tau)\, d\tau = V, \tag{4.114}$$

or

$$\frac{d^2i}{dt^2} + 2\frac{di}{dt} + 2i = 0. \tag{4.115}$$

The characteristic equation is

$$s^2 + 2s + 2 = 0, \tag{4.116}$$

which has the roots

$$s_{1,2} = -1 \pm j1.$$

The general solution is therefore

$$i(t) = K_1 \epsilon^{(-1+j1)t} + K_2 \epsilon^{(-1-j1)t}. \tag{4.117}$$

Putting in the initial conditions,

$$i(0+) = K_1 + K_2 = 0 \tag{4.118}$$

$$\left.\frac{di}{dt}\right|_{0+} = (-1 + j1)K_1 + (-1 - j1)K_2 = V. \tag{4.119}$$

Thus,

$$K_1 = -K_2 = \frac{V}{2j}.$$

The complete solution is, therefore,

$$i(t) = \frac{V}{2j}[\epsilon^{(-1+j1)t} - \epsilon^{(-1-j1)t}]$$

$$= V\epsilon^{-t}\frac{\epsilon^{jt} - \epsilon^{-jt}}{2j} = V\epsilon^{-t}\sin t. \tag{4.120}$$

This solution is sketched in Figure 4.32.

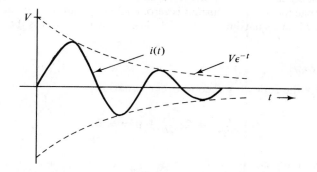

**Figure 4.32** Sketch of Solution of Example 4.20

### Example 4.21

We will find the frequency of oscillation of the circuit of Figure 4.33. The mesh equation is

$$L\frac{di}{dt} + \frac{1}{C}\int_{-\infty}^{t} i(\tau)\, d\tau = 0, \tag{4.121}$$

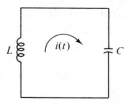

**Figure 4.33** Circuit for Example 4.21

or differentiating,

$$L\frac{d^2i}{dt^2} + \frac{1}{C}i = 0. \tag{4.122}$$

The characteristic equation is

$$Ls^2 + \frac{1}{C} = 0, \qquad s^2 + \frac{1}{LC} = 0, \tag{4.123}$$

and the roots are $s_{1,2} = \pm j(1/\sqrt{LC})$. Therefore, the general solution has the form

$$i(t) = K_1\epsilon^{+j(1/\sqrt{LC})t} + K_2\epsilon^{-j(1/\sqrt{LC})t}. \tag{4.124}$$

Since there is no real exponential part of the solution this time, the solution is a steady-state sinusoid, of radian frequency $\omega = 1/\sqrt{LC}$. Note that since there is no resistance in the circuit, no energy is dissipated, and the energy is alternately stored in the inductor and capacitor, being entirely stored in the capacitor when the current is zero, and entirely in the inductor when the current is at its maximum, at which times the voltage across the capacitor is zero.

### Repeated Roots

Consider the following second-order equation:

$$\frac{d^2x}{dt^2} + 2a\frac{dx}{dt} + a^2x = 0. \tag{4.125}$$

The characteristic equation is

$$s^2 + 2as + a^2 = 0, \tag{4.126}$$

which has the two identical roots $s_1 = s_2 = -a$.

Proceeding as before, we would write our general solution as

$$x(t) = K_1\epsilon^{-at} + K_2\epsilon^{-at} = K_3\epsilon^{-at}, \tag{4.127}$$

where $K_3 = K_1 + K_2$. Thus there is only a single constant—but to satisfy initial conditions a second-order equation must have two arbitrary constants. We are therefore led to look for another term in the solution. Since the exponential is so fundamental in the solution of our differential equations, we might look for another solution by modifying the exponential—in particular, let us try multiplying it by another time function, say $g(t)$. We will therefore try a solution of the form

$$x(t) = g(t)\epsilon^{-at}. \tag{4.128}$$

Substituting this into equation 4.125 yields the condition

$$\frac{d^2 g(t)}{dt^2} = 0,$$
(4.129)

which, upon integrating twice, gives

$$g(t) = K_1 + K_2 t.$$
(4.130)

Our solution is therefore

$$x(t) = K_1 \epsilon^{-at} + K_2 t \epsilon^{-at},$$
(4.131)

as may be verified by substitution into equation 4.125.

An identical procedure may be used for a third-order equation with three repeated roots; in this case the trial solution (equation 4.128) yields the condition

$$\frac{d^3 g(t)}{dt^3} = 0,$$
(4.132)

or

$$g(t) = K_1 + K_2 t + K_3 t^2,$$
(4.133)

and

$$x(t) = K_1 \epsilon^{-at} + K_2 t \epsilon^{-at} + K_3 t^2 \epsilon^{-at},$$
(4.134)

and the extension to higher-order repeated roots is obvious.

Of course, the number of repeated roots may be less than the order of the equation—for example, a third-order equation may have only two identical roots. In this case the distinct roots have associated with them the ordinary exponential terms, with the special forms of equations 4.131 and 4.134 associated only with the repeated roots.

### Root Positions in the Complex Plane

We have seen that the solution of an ordinary linear homogeneous differential equation consists basically of exponential terms, with the exponents given by the roots of the characteristic equation. This characteristic equation is an algebraic equation with real coefficients, and the theory of equations tells us that there are certain constraints on the roots of such an equation. Some of these constraints are as follows:

1. If the roots are imaginary or complex, they must occur as conjugate pairs.
2. If the order of the characteristic equation is odd, at least one root must be real, with the remaining roots either real or in conjugate pairs.

In addition, if a physical system is stable, no growing exponentials are allowed in the solution—this means that all roots must have nonpositive real parts. (The subject of stability is considered in some detail in Chapter 8.) Furthermore, we know that our solutions to physical problems must be real, and this implies that the coefficients that multiply conjugate imaginary or complex exponential terms in the solution must themselves be conjugate pairs, when the initial conditions are substituted, as indicated in equation 4.38.

It is helpful to visualize the roots of the characteristic equation as points in the complex plane, or s-plane. Figure 4.34 indicates the complex plane, with three real roots, a pair of complex roots, and a pair of imaginary roots

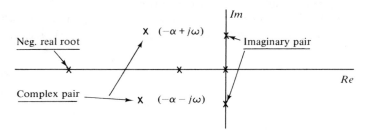

**Figure 4.34**   Root Positions in the Complex Plane

plotted. For the reason given previously, all the roots are shown with nonpositive real parts, and as conjugate pairs if they are imaginary or complex. Each root contributes an exponential term to the time response, in the following way:

1. Each real root at a position $(-\alpha)$ contributes a real exponential term $K\epsilon^{-\alpha t}$. (The root at the origin contributes a constant, $K\epsilon^{-0t} = K$.)
2. Each conjugate complex pair (along with their conjugate coefficients) contributes an exponentially damped sinusoidal term to the solution; i.e., a term of the form given by equation 4.39 or equation 4.41. The speed of the exponential decay is proportional to the distance of the roots from the imaginary axis $(-\alpha)$, while the frequency of the sinusoidal part of the term is proportional to the distance of the roots from the real axis $(\omega)$.
3. Each conjugate imaginary pair contributes a steady-state sinusoidal term to the solution, the frequency of which is proportional to the distance of the roots from the real axis. $(\alpha = 0$, so there is no exponential decay factor.)

Thus the total solution consists of real exponential, sinusoidal, or damped sinusoidal terms, the basic characteristics of which are determined by the positions of each real root and each conjugate pair of complex or imaginary roots in the complex plane. (In the case of repeated roots, terms involving real exponentials multiplied by powers of $t$ also occur.) Of course, the root positions do not tell everything about the solution—the constants associated with each term are not given by the root positions but must be evaluated from the initial conditions, and these constants specify how much of each characteristic exponential or sinusoidal response is present for a particular set of initial conditions. In some instances the solution term associated with a root (or pair of roots) may have such a small relative amplitude that it may be neglected; in some cases, only one root will dominate the solution—but this information is not contained in the root-position diagram.

### Root Positions for a Second-Order Network

Let us consider a second-order homogeneous differential equation of the form

$$\frac{d^2x}{dt^2} + a\frac{dx}{dt} + bx = 0. \tag{4.135}$$

The associated characteristic equation is

$$s^2 + as + b = 0. \tag{4.136}$$

It is common practice, particularly by control engineers, to write this equation in the form

$$s^2 + 2\zeta\omega_0 s + \omega_0^2 = 0; \tag{4.137}$$

i.e., we have defined

$$\omega_0 \triangleq \sqrt{b}, \qquad \zeta \triangleq \frac{a}{2\sqrt{b}} \tag{4.138}$$

The roots of the characteristic equation, when it is written in the form of equation 4.137, are

$$s_{1,2} = -\zeta\omega_0 \pm \omega_0\sqrt{\zeta^2 - 1}$$
$$= -\zeta\omega_0 \pm j\omega_0\sqrt{1 - \zeta^2}. \tag{4.139}$$

Examining these roots, we note that:

1. For $\zeta = 0$, the roots are pure imaginary ($s_{1,2} = \pm j\omega_0$). The solution has the form of a sinusoid of constant amplitude, and the system or circuit is said to be *undamped*.
2. For $0 < \zeta < 1$, the roots are a conjugate complex pair. The solution has the form of an exponentially damped sinusoid, and the system is said to be *underdamped*.

3. If $\zeta = 1$, the roots are identical ($s_{1,2} = -\omega_0$) and real. This is the value of $\zeta$ at which the solution becomes no longer oscillatory, and the system is said to be *critically damped*.
4. For $\zeta > 1$, the roots are distinct and negative reals. The solution consists of the sum of two decaying real exponentials, and the system is said to be *overdamped*.

The term $\zeta$ is usually called the *damping factor*, and, since with $\zeta = 0$ the solution is a nondecaying oscillation of radian frequency $\omega_0$, $\omega_0$ is usually called the *undamped natural frequency* of the circuit or system.

Figure 4.35 shows the locus of the roots of the second-order characteristic equation in the $s$-plane, as $\zeta$ is varied from zero to infinity. For $0 < \zeta < 1$, the locus is a circular arc, of radius

$$\sqrt{(\zeta\omega_0)^2 + (\omega_0\sqrt{1 - \zeta^2})^2} = \omega_0 \tag{4.140}$$

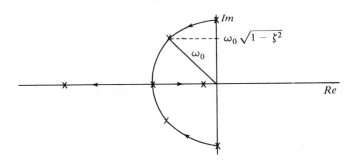

**Figure 4.35** Locus of Roots as $\zeta$ Is Varied

At $\zeta = 1$, the roots come together on the negative real axis, and for $\zeta > 1$ they split apart on the negative real axis, the one approaching zero as $\zeta \to \infty$ and the other approaching infinity as $\zeta \to \infty$, as is readily seen by examining equation 4.139. Note that as $\zeta$ is increased from zero, the oscillation in the solution not only becomes exponentially damped, but its frequency is also decreased by the factor $\sqrt{1 - \zeta^2}$, and for $0 < \zeta < 1$ the solution may be written in the form

$$x(t) = K\epsilon^{-\zeta\omega_0 t} \cos [(\omega_0\sqrt{1 - \zeta^2})t + \theta], \tag{4.141}$$

where $K$ and $\theta$ are determined by the initial conditions. The oscillatory part of this solution has a frequency of $\omega_0\sqrt{1 - \zeta^2}$ radians/second, and this is often referred to as the *damped natural frequency* of the circuit or system.

**Example 4.22**

Consider the series $RLC$ circuit of Figure 4.36. The mesh equation is

$$L \frac{di}{dt} + Ri + \frac{1}{C} \int_{-\infty}^{t} i(\tau)\, d\tau = 0, \tag{4.142}$$

or

$$\frac{d^2i}{dt} + \frac{R}{L} \frac{di}{dt} + \frac{1}{LC} i = 0. \tag{4.143}$$

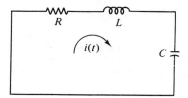

**Figure 4.36**  Series $R, L, C$ Circuit of Example 4.22

By comparison with equation 4.135, it is clear that in this example

$$\omega_0 = \sqrt{\frac{1}{LC}} \quad \text{and} \quad \zeta = \frac{R}{2} \sqrt{\frac{C}{L}},$$

and the solution may be written as

$$i(t) = K \epsilon^{-(R/2L)t} \cos\left( \sqrt{\frac{1}{LC} - \frac{R^2}{4L^2}}\, t + \theta \right), \tag{4.144}$$

where again $K$ and $\theta$ must be determined from initial conditions [i.e., initial inductor current and capacitor voltage, from which $i(0)$ and $di/dt|_0$ can be found].

Note that in this circuit, if $R$ is zero, the solution is a steady-state oscillation of frequency $\sqrt{1/LC}$ radians/second. Also note that the damping factor is directly proportional to the resistance in the circuit, if $L$ and $C$ are fixed. Critical damping ($\zeta = 1$) occurs at a resistance value of

$$R_{\text{crit}} = 2 \sqrt{\frac{L}{C}}, \tag{4.145}$$

and for $R \geq R_{\text{crit}}$, the solution is nonoscillatory.

## PROBLEMS

**4.1**  In the network of Figure 4.37, find the constant-steady-state current in each branch of the network.

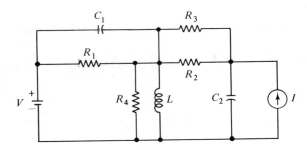

**Figure 4.37** Network in the Constant Steady State

**4.2** In the circuit of Figure 4.38, the capacitor voltage at $t = 0-$ is 1 V, as shown. At $t = 0$ the switch is closed. Find the values of the current and its first two derivatives at $t = 0+$, and find $i(t)$ for $t > 0$.

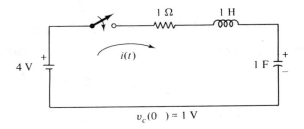

**Figure 4.38** Circuit of Problem 4.2

**4.3** Repeat Problem 4.2 for a resistance value of 2 Ω in the circuit.

**4.4** Repeat Problem 4.2 for a resistance value of 4 Ω in the circuit.

**4.5** The circuit of Figure 4.39 is in the constant steady state prior to $t = 0$, and at $t = 0$ the switch is thrown in the direction indicated by the arrow. Find the values of $v(t)$ and its first two derivatives at $t = 0+$, and find $v(t)$ for $t > 0$.

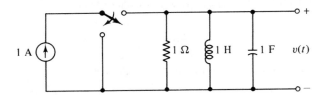

**Figure 4.39** Circuit of Problem 4.5

**4.6** In the circuit of Figure 4.37, the current source is removed at time $t = 0$, the circuit previously being in the constant steady state. Draw an equivalent circuit valid for $t = 0+$, and find each branch current at $t = 0+$.

**4.7** Write expressions for the time functions sketched in Figure 4.40, using step-function notation.

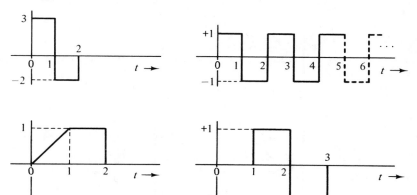

**Figure 4.40** Some Waveforms Related to Step Functions

**4.8** Write expressions for the pulse waveforms shown in Figure 4.41.

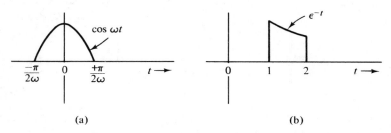

(a)                               (b)

**Figure 4.41** Pulse Waveforms of Problem 4.8

**4.9** In the networks shown in Figure 4.42, the switches are thrown in the directions indicated by the arrows at $t = 0$, the networks being in the constant steady state prior to $t = 0$. In each case, find the initial and final values of the time functions indicated, and the time constant. Sketch each quantity as a function of time for $t > 0$.

**4.10** Write a mathematical expression for each of the functions sketched in Problem 4.9, using the general form of first-order solution as given in equation 4.87.

**4.11** Find the solutions for the following second-order differential equations, with the initial conditions indicated.

(a) $\dfrac{d^2x}{dt^2} + 2\dfrac{dx}{dt} + 2x = 0,$     $x(0) = 1,$     $\left.\dfrac{dx}{dt}\right|_0 = 1.$

(b) $\dfrac{d^2x}{dt^2} + 4x = 0,$     $x(0) = 1,$     $\left.\dfrac{dx}{dt}\right|_0 = 1.$

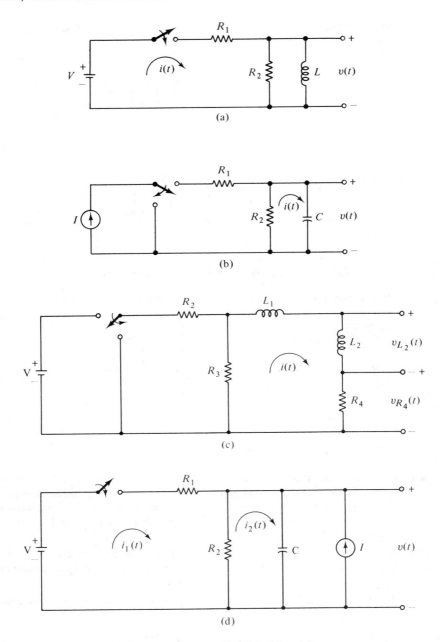

**Figure 4.42**  Networks of Problem 4.9

(c) $\dfrac{d^2x}{dt^2} + 2\dfrac{dx}{dt} + x = 0$, $\qquad x(0) = 1$, $\qquad \left.\dfrac{dx}{dt}\right|_0 = 0$.

(d) $\dfrac{dx}{dt} + x + \displaystyle\int_{-\infty}^{t} x(\tau)\,d\tau = 5$, $\qquad x(0) = 0$, $\qquad \left.\dfrac{dx}{dt}\right|_0 = 1$.

**4.12**  Discuss carefully the relationship of the root positions of the characteristic equation roots in the complex plane and the time response of the network.

**4.13**  Figure 4.43 shows the root positions of a certain characteristic equation. Write the corresponding time solution (in terms of arbitrary constants).

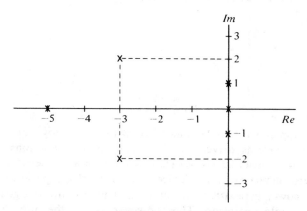

**Figure 4.43**  Root Positions for Problem 4.13

**4.14**  Show that the trial solution of equation 4.128 results in the condition of equation 4.129, and that this in turn leads to equation 4.131.

**4.15**  Find the damping ratio ($\zeta$) and the undamped natural frequency ($\omega_0$) for the parallel $RLC$ network of Figure 4.39. For what value of $R$ will the network be critically damped?

**4.16**  Show that, if a linear differential equation is satisfied by a complex time function (such as a complex exponential), then both the real and imaginary parts of the time function will also satisfy the equation. (This fact will be utilized in Chapter 5.)

# 5

# Systems with
# Harmonic Forcing Functions

In this chapter we will be concerned with the steady-state behavior of networks and systems driven by sinusoidal forcing functions. An understanding of the sinusoidal steady state is important for several reasons. In the first place, many electrical devices and networks, especially those associated with power generation, distribution, and use, are designed for sinusoidal steady-state operation. This, however, is not the only reason to be concerned with sinusoidal signals. As we shall see later (Chapter 6), a large class of nonsinusoidal signals can be represented in terms of linear combinations of sinusoids, through the use of the Fourier series or the Fourier integral. Thus, an understanding of the behavior of a system with sinusoidal forcing functions of different frequencies leads to an understanding of the behavior of the system with other types of forcing functions. (To cite an everyday example, a high-fidelity amplifier is not supplied to the customer with test results based on a recording of Beethoven's Ninth Symphony; it is supplied to the customer with an indication of how well it amplifies steady-state sinusoids of different frequencies—i.e., a frequency-response plot.) Also, many fundamental properties (such as resonance) exhibited by linear systems are basically associated with sinusoidal signals. Thus sinusoids occupy preeminent positions as steady-state forcing functions, and it is most fortunate that linear systems in the sinusoidal steady state can be analyzed with relative ease.

If the forcing function of a linear differential equation is a sinusoid, then, as we shall see, the solution of the equation may be obtained by a relatively simple manipulation of complex numbers. Before proceeding, the reader should be thoroughly familiar with the elements of complex algebra given in Appendix A.

## 5.1 SOLUTION OF DIFFERENTIAL EQUATIONS WITH SINUSOIDAL FORCING FUNCTIONS

### General Input–Output Relationship

Let us consider an ordinary linear differential equation relating a network output or response to an input or driving function. In its most general form, this equation will relate the response function and its derivatives of various orders to the input function and its derivatives of various orders, as indicated by equation 5.1:

$$a_n D^n y(t) + a_{n-1} D^{n-1} y(t) + \ldots + a_0 y(t)$$
$$= b_m D^m x(t) + b_{m-1} D^{m-1} x(t) + \ldots + b_0 x(t). \qquad (5.1)$$

In this equation we have used the operator $D^n$ to denote the $n$th-order derivative, as we did in Chapter 4 (Example 4.19). $x(t)$ is the input function and $y(t)$ is the response; if more than one differential equation was involved in the original problem formulation, then it is assumed that the set of equations have been reduced to a single equation relating the desired input and output variables, as in Example 4.19.

### Example 5.1

The mesh current in the circuit of Figure 5.1 is described by the equation (in operator form)

$$L D i(t) + R i(t) + \frac{1}{C} D^{-1} i(t) = v(t). \qquad (5.2)$$

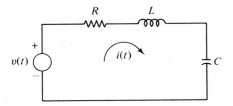

**Figure 5.1** Circuit for Example 5.1

Differentiating this equation once yields

$$\left( L D^2 + R D + \frac{1}{C} \right) i(t) = D v(t), \qquad (5.3)$$

which is of the general form of equation 5.1.

### Example 5.2

All element values in the network of Figure 5.2 are unity. Find a differential equation relating $v_2(t)$ to $i(t)$.

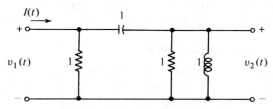

**Figure 5.2** Circuit for Example 5.2

Writing equations for the two unknown node voltages (using Kirchhoff's current law) yields the two equations

$$D[v_1(t) - v_2(t)] + v_1(t) = i(t)$$
$$D[v_1(t) - v_2(t)] - v_2(t) - D^{-1}v_2(t) = 0. \tag{5.4}$$

Multiplying the second equation by $D$ (differentiating) and rearranging terms yields

$$(D + 1)v_1(t) - Dv_2(t) = i(t)$$
$$D^2v_1(t) - (D^2 + D + 1)v_2(t) = 0. \tag{5.5}$$

Using Cramer's rule to solve for $v_2(t)$, we have

$$v_2(t) = \frac{\begin{vmatrix} D + 1 & i(t) \\ D^2 & 0 \end{vmatrix}}{\begin{vmatrix} D + 1 & -D \\ D^2 & -(D^2 + D + 1) \end{vmatrix}}$$

$$= \frac{D^2 i(t)}{2D^2 + 2D + 1}. \tag{5.6}$$

Therefore, the required differential equation is

$$(2D^2 + 2D + 1)v_2(t) = D^2 i(t), \tag{5.7}$$

or

$$2\frac{d^2 v_2(t)}{dt^2} + 2\frac{dv_2(t)}{dt} + v_2(t) = \frac{d^2 i(t)}{dt^2}. \tag{5.8}$$

### Output Response for Sinusoidal Inputs

Consider a forcing function of the form

$$f(t) = C\epsilon^{j\omega t} = (a + jb)\epsilon^{j\omega t}, \tag{5.9}$$

where $C$ is the complex number $a + jb$. This is a complex time function, and by using Euler's relationships we may express it in the form

$$f(t) = (a + jb)(\cos \omega t + j \sin \omega t)$$
$$= (a \cos \omega t - b \sin \omega t) + j(a \sin \omega t + b \cos \omega t). \quad (5.10)$$

Thus the *real* part of $f(t)$ is simply

$$\text{Re}[f(t)] = a \cos \omega t - b \sin \omega t$$
$$= \sqrt{a^2 + b^2} \cos \left( \omega t + \tan^{-1} \frac{b}{a} \right)$$
$$= |C| \cos (\omega t + /C), \quad (5.11)$$

where

$$|C| \triangleq \sqrt{a^2 + b^2} \quad \text{and} \quad \underline{/C} \triangleq \tan^{-1} \frac{b}{a}.$$

Hence the real part of $f(t)$ is simply a general sinusoidal function; i.e., a sinusoid whose determining characteristics (amplitude, phase, and frequency) may be freely specified by a choice of the three constants, $a$, $b$, and $\omega$. The amplitude of the sinusoid is $|C|$, its phase angle is $\underline{/C}$ (relative to cos $\omega t$), and its frequency is $\omega$ radians/second, or $\omega/2\pi$ cycles/second (hertz). This is the type of driving function in which we are now interested. Rather than using $\text{Re}[f(t)]$ directly, however, we will see that there is some advantage to using the complex function $f(t)$ itself as a forcing function for our equations. Since our equations are linear, both the real and imaginary parts of any complex solution must themselves be solutions (Problem 4.16); hence we may look for a solution using $f(t)$ as the forcing function, and we are assured that the real part of the solution obtained will be the solution of the same equation with $\text{Re}[f(t)]$ as the forcing function.

Let us return now to the general differential equation of the form of equation 5.1, which we may rewrite as

$$(a_n D^n + a_{n-1} D^{n-1} + \ldots + a_0)y(t)$$
$$= (b_m D^m + b_{m-1} D^{m-1} + \ldots + b_0)x(t). \quad (5.12)$$

Since both the response function $y(t)$ and the forcing function $x(t)$ are multiplied by polynomials in the operator $D$, we may abbreviate this equation by writing

$$M(D)y(t) = N(D)x(t), \quad (5.13)$$

where $M(D)$ and $N(D)$ are polynomials in $D$. Since

$$D^n(C\epsilon^{j\omega t}) \triangleq \frac{d^n}{dt^n}(C\epsilon^{j\omega t})$$
$$= (j\omega)^n C\epsilon^{j\omega t}, \quad (5.14)$$

it is clear that letting $x(t) = C\epsilon^{j\omega t}$ allows us to write equation 5.13 as

$$M(D)y(t) = N(j\omega)C\epsilon^{j\omega t}. \tag{5.15}$$

Since this equation must be satisfied at every instant of time, the logical choice for a trial solution is another complex exponential of the same general form as $f(t)$; i.e.,

$$y(t) = K\epsilon^{j\omega t} = (c + jd)\epsilon^{j\omega t}. \tag{5.16}$$

Substituting this trial solution into equation 5.15 yields

$$M(j\omega)K = N(j\omega)C, \tag{5.17}$$

so the trial solution works provided that

$$K = \frac{N(j\omega)}{M(j\omega)}C. \tag{5.18}$$

Thus, if we let $H(j\omega) \triangleq N(j\omega)/M(j\omega)$, then

$$|K| = |H(j\omega)||C| \tag{5.19}$$

and

$$\underline{/K} = \underline{/H(j\omega)} + \underline{/C}. \tag{5.20}$$

Since the real part of $y(t)$ must be the solution to the equation if our actual forcing function is the real part of $x(t)$, it follows that for the forcing function

$$x(t) = \text{Re}(C\epsilon^{j\omega t}) = |C|\cos(\omega t + \underline{/C}), \tag{5.21}$$

the solution is

$$y(t) = \text{Re}(K\epsilon^{j\omega t}) = |K|\cos(\omega t + \underline{/K}), \tag{5.22}$$

where the magnitudes and angles of $C$ and $K$ are related by (5.19) and (5.20). We may therefore make the following statement: The ratio of the amplitude (peak value) of the output sinusoid to that of the input sinusoid is $|H(j\omega)|$, and the phase difference between the output and input sinusoids is $\underline{/H(j\omega)}$.

In a general sense, this statement summarizes everything that there is to be known about the behavior of lumped, linear, time-invariant systems in the sinusoidal steady state. It implies that every quantity in the circuit or system has a sinusoidal time variation with the same frequency as the forcing function sinusoid. Each quantity can differ from the forcing function only in amplitude and phase angle. To satisfy the differential equation we needed a function which, when differentiated any number of times, preserved its general functional form, and the only periodic function having this property is the sinusoid. It is this property which endows the sinusoid with such importance and makes it preeminent within the class of periodic forcing functions.

### Example 5.3

In the circuit of Figure 5.3, $v_1(t)$ is a steady-state sinusoid having a frequency of $\omega$ radians/second. Find the relative amplitude and phase angle of the output sinusoid, $v_2(t)$.

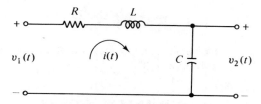

**Figure 5.3**   Circuit of Example 5.3

Writing the differential equations for this circuit in operator form, we have

$$\left(LD + R + \frac{1}{CD}\right)i(t) = v_1(t) \tag{5.23}$$

$$CDv_2(t) = i(t), \tag{5.24}$$

which may be combined to give

$$(LCD_2 + RCD + 1)v_2(t) = v_1(t). \tag{5.25}$$

Thus, for this example,

$$M(D) = LCD^2 + RCD + 1$$

and

$$N(D) = 1;$$

therefore,

$$H(j\omega) \triangleq \frac{N(j\omega)}{M(j\omega)} = \frac{1}{-\omega^2 LC + j\omega RC + 1}$$

$$= \frac{1}{(1 - \omega^2 LC) + j(\omega RC)}. \tag{5.26}$$

Hence the ratio of the amplitudes (peak values) of $v_2(t)$ and $v_1(t)$ is

$$|H(j\omega)| = \frac{|1|}{|(1 - \omega^2 LC) + j(RC)|}$$

$$= \frac{1}{\sqrt{(1 - \omega^2 LC)^2 + (\omega RC)^2}}. \tag{5.27}$$

Also, $v_2(t)$ is shifted in phase from $v_1(t)$ by the angle

$$\underline{/H(j\omega)} = \underline{/1} - \underline{/(1 - \omega^2 LC) + j(\omega RC)}$$
$$= -\tan^{-1} \frac{\omega RC}{1 - \omega^2 LC} . \tag{5.28}$$

If, for example, we specify that $v_1(t) = \cos \omega t$, then

$$v_2(t) = \frac{1}{\sqrt{(1 - \omega^2 LC)^2 + (\omega RC)^2}} \cos \left( \omega t - \tan^{-1} \frac{\omega RC}{1 - \omega^2 LC} \right), \tag{5.29}$$

and, in fact, we may immediately write down the expression for $v_2(t)$ for *any* specific sinusoidal form of $v_1(t)$.

## 5.2  PHASOR ANALYSIS OF CIRCUITS

### *Phasor Representation of Sinusoids*

A phasor diagram is a graphical representation, in the complex plane, of complex exponentials. Let us first consider the function

$$\epsilon^{j\omega t} = \cos \omega t + j \sin \omega t \tag{5.30}$$

and describe its locus in the complex plane. We may consider any complex number to be a point in the complex plane, or, equivalently, a vector drawn from the origin to that point, as indicated in Figure 5.4. The magnitude and angle of a complex number, as previously defined (equation 5.11), represent the length of the vector, and its angle measured from the positive real axis,

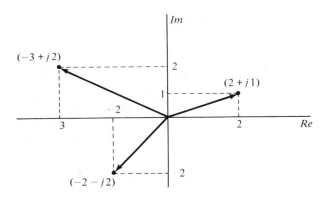

**Figure 5.4**  The Complex Plane

respectively. The function $\epsilon^{j\omega t}$ is, of course, just a complex number for any specified value of $t$. Its magnitude is

$$|\epsilon^{j\omega t}| = \sqrt{\cos^2 \omega t + \sin^2 \omega t} = 1 \qquad (5.31)$$

and its angle is

$$\underline{/\epsilon^{j\omega t}} = \tan^{-1} \frac{\sin \omega t}{\cos \omega t} = \omega t. \qquad (5.32)$$

In the complex plane, $\epsilon^{j\omega t}$ is therefore represented by a vector of unit length and angle $\omega t$; it rotates around the origin at the constant angular rate of $\omega$ radians/second, and lies along the positive real axis at the epochs given by $t = 0,\ \pm 2\pi/\omega,\ \pm 4\pi/\omega, \ldots, \pm m\pi/\omega$. This rotating vector in the complex plane is termed a phasor. (The reason for using this term, rather than vector, is that many physical problems involve spatial vector quantities that may have sinusoidal time variations, so that two different types of vector quantities are involved at once, which can be very confusing unless a special term [phasor] is used to represent the sinusoidal time variation.) The phasor representing $\epsilon^{j\omega t}$ is shown in Figure 5.5.

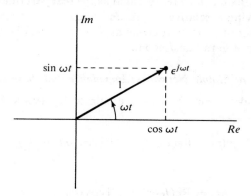

**Figure 5.5**  Phasor Representation of $\epsilon^{j\omega t}$

Now let us consider the phasor representation for

$$C\epsilon^{j\omega t} = (a + jb)\epsilon^{j\omega t}. \qquad (5.33)$$

This phasor has a length

$$|C\epsilon^{j\omega t}| = |C||\epsilon^{j\omega t}| = |C| = \sqrt{a^2 + b^2} \qquad (5.34)$$

and an angle

$$\underline{/C\epsilon^{j\omega t}} = \underline{/C} + \underline{/\epsilon^{j\omega t}} = \omega t + \underline{/C}. \qquad (5.35)$$

Thus it rotates at the same angular rate ($\omega$) as does $\epsilon^{j\omega t}$, but it has a different length and angle, as shown in Figure 5.6. The lengths and angular difference

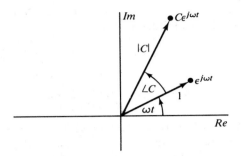

**Figure 5.6**  Phasor $C\epsilon^{j\omega t}$

between these two phasors do not change with time; the two phasors rotate together, as if locked in position with respect to each other, at the angular rate $\omega$. Since, in a linear circuit driven by a sinusoidal forcing function, all voltages and currents (as well as charges and flux linkages) are sinusoids differing from each other only in amplitude and phase, it follows that each quantity in the circuit can be represented as the real part of a rotating phasor, each phasor having a relative length and angle with respect to the others, with the entire set of phasors rotating as if fastened rigidly together, with the angular velocity of $\omega$ radians/second.

### *Phasor Analysis of Simple Networks: Impedance and Admittance*

Let us consider a circuit with one pair of terminals, for which we can define a voltage

$$v(t) = \text{Re}(V\epsilon^{j\omega t}) = |V|\cos(\omega t + \underline{/V}), \tag{5.36}$$

and a current

$$i(t) = \text{Re}(I\epsilon^{j\omega t}) = |I|\cos(\omega t + \underline{/I}). \tag{5.37}$$

At $t = 0, \pm 2\pi/\omega, \pm 4\pi/\omega, \ldots$, the voltage and current phasors are simply specified by the complex numbers $V$ and $I$, respectively. In general, the ratio of these two complex coefficients will depend upon $\omega$, in the same way that $C$ and $K$ are related in equations 5.19 and 5.20. Now, however, we will use symbols other than $H(j\omega)$ to describe the ratios of $V$ to $I$ or $I$ to $V$, and we will call these ratios impedance and admittance, respectively; i.e.,

$$Z(j\omega) \triangleq \frac{V}{I} \qquad \text{(impedance)} \tag{5.38}$$

$$Y(j\omega) \triangleq \frac{I}{V} = \frac{1}{Z(j\omega)} \qquad \text{(admittance).} \tag{5.39}$$

The impedance and admittance are thus complex numbers themselves, and they describe how the voltage and current phasors are related. That is,

$$\frac{|V|}{|I|} = |Z(j\omega)| \tag{5.40}$$

$$\underline{/V} - \underline{/I} = \underline{/Z(j\omega)} \tag{5.41}$$

$$\frac{|I|}{|V|} = |Y(j\omega)| \tag{5.42}$$

$$\underline{/I} - \underline{/V} = \underline{/Y(j\omega)}. \tag{5.43}$$

For individual resistors, inductors, and capacitors, the impedance and admittance are given in Table 5.1. As indicated in the table, the voltage *leads* the current by 90° in an inductor and *lags* the current by 90° in a capacitor. For a resistor, of course, the voltage and current are in phase.

**Table 5.1** Impedance Relationships for Single Elements

| Element | $Z(j\omega)$ | $Y(j\omega)$ | $\dfrac{\|V\|}{\|I\|}$ | $\underline{/V} - \underline{/I}$ |
|:---:|:---:|:---:|:---:|:---:|
| $R$ | $R$ | $\dfrac{1}{R}$ | $R$ | $0$ |
| $L$ | $j\omega L$ | $\dfrac{1}{j\omega L}$ | $\omega L$ | $+\dfrac{\pi}{2}$ |
| $C$ | $\dfrac{1}{j\omega C}$ | $j\omega C$ | $\dfrac{1}{\omega C}$ | $-\dfrac{\pi}{2}$ |

If elements are connected in series, then the current phasor is common, and the voltage phasor across the series combination is just the sum of the element voltage phasors (Kirchhoff's voltage law); hence impedances in series add, just as in the case of resistors in series. For impedances in parallel, the voltage phasor is common and the current phasors add—thus the admittance of the parallel combination is the sum of the individual admittances. Hence we may write

$$\text{For series elements:} \quad Z_{\text{tot}}(j\omega) = \sum_i Z_i(j\omega) \tag{5.44}$$

$$\text{For parallel elements:} \quad Y_{\text{tot}}(j\omega) = \sum_i Y_i(j\omega). \tag{5.45}$$

(Actually, the same rules of combination hold if the individual $Z_i$'s and $Y_i$'s are the impedances and admittances of multi-element circuits themselves, rather than single elements; why?) The rules for combining impedances and admittances may be used to simplify the sinusoidal-steady-state analysis of networks, as indicated in the following examples.

### Example 5.4

We will use impedances to analyze the circuit of Figure 5.3. Letting

$$v_1(t) = \text{Re}(V_1 \epsilon^{j\omega t})$$
$$v_2(t) = \text{Re}(V_2 \epsilon^{j\omega t})$$

and

$$i(t) = \text{Re}(I \epsilon^{j\omega t}),$$

we have

$$I = \frac{V_1}{Z_1(j\omega)} = \frac{V_1}{R + j\omega L + (1/j\omega C)} = \frac{j\omega C V_1}{(1 - \omega^2 LC) + j\omega RC}, \quad (5.46)$$

where $Z_1(j\omega) = R + j\omega L + (1/j\omega C)$ is the *input impedance*, $V_1/I$. Then

$$V_2 = I \frac{1}{j\omega C} = \frac{V_1}{(1 - \omega^2 LC) + j\omega RC}, \quad (5.47)$$

or

$$\frac{V_2}{V_1} = H(j\omega) = \frac{1}{(1 - \omega^2 LC) + j\omega RC}, \quad (5.48)$$

which is identical to the result obtained previously in a more lengthy derivation. $V_2/V_1$ is one type of *transfer function*, called a *voltage transfer ratio*. (Impedances and admittances themselves are also transfer functions; this term is used in quite a broad sense to indicate the ratio of "output" to "input" phasors, whether these phasors are voltages, currents, forces, velocities, charges, or any other variables.)

### Example 5.5

Figure 5.7 shows a general configuration called a *voltage divider*, of which the circuit of Figure 5.3 is a special case. Since $V_2$ is $Z_2(j\omega)$ multiplied by the current phasor, we have

$$V_2 = Z_2(j\omega)I = \frac{Z_2(j\omega)V_1}{Z_1(j\omega) + Z_2(j\omega)}, \quad (5.49)$$

or

$$\frac{V_2}{V_1} = \frac{Z_2(j\omega)}{Z_1(j\omega) + Z_2(j\omega)}. \quad (5.50)$$

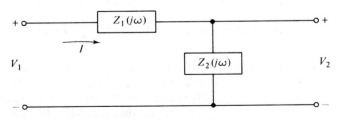

**Figure 5.7**   Voltage Divider

Thus the ratio of $V_2$ to $V_1$ is the same as the impedance that $V_2$ appears across divided by the impedance that $V_1$ appears across. Applying this to the previous example, the voltage transfer ratio of equation 5.48 may be written immediately.

### Example 5.6

A very common form of network is a ladder configuration, such as the one indicated in Figure 5.8. Two series and two shunt branches appear in this ladder, but the ladder may be extended to include as many series impedances and shunt admittances as desired. (Each additional "rung" will provide a new mesh.) Again, each impedance and admittance may consist of a single element, or each may be a complex circuit in itself.

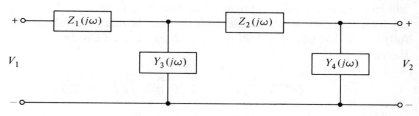

**Figure 5.8** Ladder Configuration

Let us suppose that we desire to find the voltage transfer ratio, $V_2/V_1$, for this circuit. The easiest way to proceed, for such a ladder configuration, is to start at the output end and work backward toward the input end. Thus

$$
\begin{aligned}
I_{y4} &= V_2 Y_4 & I_{y3} &= V_{y3} Y_3 \\
I_{z2} &= I_{y4} & I_{z1} &= I_{y3} + I_{z2} \\
V_{z2} &= I_{z2} Z_2 & V_{z1} &= I_{z1} Z_1 \\
V_{y3} &= V_2 + V_{z2} & V_1 &= V_{z1} + V_{y3}.
\end{aligned}
\tag{5.51}
$$

We have now calculated $V_1$ in terms of $V_2$, and we may write the desired ratio, $V_2/V_1$. Some specific ladder network examples will be found in the problems at the end of the chapter.

### Reactance and Susceptance

Impedance and admittance are complex functions of $\omega$; i.e., for any specified value of $\omega$, $Z(j\omega)$ and $Y(j\omega)$ are complex numbers, with real and imaginary parts. We may write

$$
Z(j\omega) = R(j\omega) + jX(j\omega)
\tag{5.52}
$$

and

$$
Y(j\omega) = G(j\omega) + jB(j\omega),
\tag{5.53}
$$

in order to show the real and imaginary parts of these functions explicitly. The terms applied to these real and imaginary components of the impedance and admittance are as follows:

$R(j\omega)$:  resistance

$X(j\omega)$:  reactance

$G(j\omega)$:  conductance

$B(j\omega)$:  susceptance.

For a single resistor, of course, $Z(j\omega) = R$ (a constant) and $Y(j\omega) = G$, where $G = 1/R$. For a more complicated circuit, however, the impedance will have a real (resistive) part, which depends upon $\omega$, as well as an imaginary (reactive) part which also depends upon $\omega$. In this case it should be carefully noted that $G(j\omega) \neq 1/R(j\omega)$, and $B(j\omega) \neq 1/X(j\omega)$. In fact (dropping the $j\omega$'s),

$$Y = \frac{1}{Z} = \frac{1}{R+jX} = \frac{1}{R+jX}\frac{R-jX}{R-jX} = \frac{R-jX}{R^2+X^2}. \tag{5.54}$$

Hence

$$G = \frac{R}{R^2+X^2} \tag{5.55}$$

and

$$B = \frac{-X}{R^2+X^2}. \tag{5.56}$$

***Example 5.7***

Figure 5.9 shows a series $RLC$ circuit with impedances of the elements indicated. The impedance appearing at the terminals is

$$Z(j\omega) = \frac{V}{I} = R + j\omega L + \frac{1}{j\omega C}$$

$$= R + j\left(\omega L - \frac{1}{\omega C}\right). \tag{5.57}$$

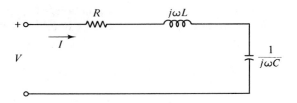

**Figure 5.9**  Series $RLC$ Circuit

In this case, the resistive part of the impedance is just the constant $R$. We might diagram this impedance in the complex plane as in Figure 5.10. The total impedance is just the vector sum of the three components, $R, j\omega L$, and $I/j\omega C$. Note that at one particular value of $\omega$,

$$\omega = \frac{1}{\sqrt{LC}}, \tag{5.58}$$

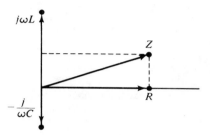

**Figure 5.10** Impedance Diagram for Circuit of Figure 5.9

$Z = R$ and the inductor and capacitor effectively "disappear" as far as the terminal characteristic of the circuit is concerned. At this frequency, $|Z(j\omega)|$ is a minimum, so for a constant voltage amplitude the current amplitude will be a maximum; this phenomena is called *series resonance*.

**Example 5.8**

Let us calculate the impedance of the parallel $RLC$ circuit of Figure 5.11. The admittance, seen from the terminals, is

$$Y(j\omega) = \frac{1}{R} + \frac{1}{j\omega L} + j\omega C$$

$$= \frac{1}{R} + j\left(\omega C - \frac{1}{\omega L}\right). \tag{5.59}$$

At the resonant frequency, $\omega = 1/\sqrt{LC}$, $Y(j\omega) = 1/R$, and again the inductor and capacitor effectively disappear as far as the terminal characteristic

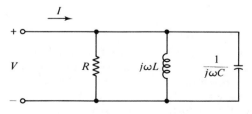

**Figure 5.11** Parallel $RLC$ Circuit

is concerned. The admittance diagram for the circuit is as shown in Figure 5.12. Since at the resonant frequency, $|Y|$ is a minimum, it follows that for a constant voltage amplitude the current amplitude is a minimum (parallel resonance).

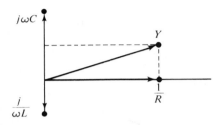

**Figure 5.12** Admittance Diagram for Circuit of Figure 5.11

To calculate the impedance, we need to invert $Y(j\omega)$; i.e.,

$$
\begin{aligned}
Z(j\omega) &= \frac{1}{Y(j\omega)} = \frac{1}{(1/R) + j[\omega C - (1/\omega L)]} \\
&= \frac{1}{(1/R) + j[\omega C - (1/\omega L)]} \frac{(1/R) - j[\omega C - (1/\omega L)]}{(1/R) - j[\omega C - (1/\omega L)]} \\
&= \frac{1/R}{(1/R)^2 + [\omega C - (1/\omega L)]^2} - j\frac{\omega C - (1/\omega L)}{(1/R)^2 + [\omega C - (1/\omega L)]^2} \\
&= (\text{resistive part}) + j(\text{reactive part}). \qquad (5.60)
\end{aligned}
$$

Note that in this circuit the resistive part of the impedance depends not only upon $R$, but upon $\omega$, $L$, and $C$ as well.

### Phasor Diagrams of Networks

As we have seen previously, when a network is operated in the sinusoidal steady state, every quantity (voltage, current, charge, flux linkage) in the circuit may be represented as the real part of a rotating phasor, and each phasor maintains a fixed magnitude and angle relationship with every other phasor. This therefore permits a graphical interpretation of Kirchhoff's laws in terms of phasor additions. This is best illustrated by means of an example.

### Example 5.9

We will draw a general phasor confirguration for the circuit of Figure 5.13. The phasor diagram may be started by drawing a phasor to represent $V_2$ in some reference position, and proceeding logically from there. For example, $I_{R_2}$ is in phase with $V_2$, $I_C$ leads $V_2$ by 90°, these two currents add to give $I$, and so forth. (Of course, the actual relative magnitudes and angles will depend upon the numerical element values and $\omega$.) The actual sinusoidal quantities

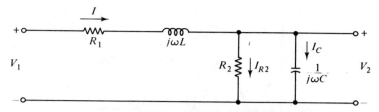

**Figure 5.13** Circuit for Example 5.9

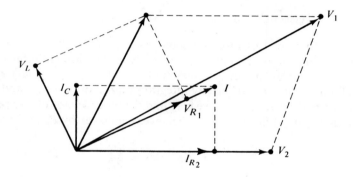

**Figure 5.14** Phasor Diagram for Circuit of Figure 5.13

in the circuit can be visualized by considering the projections of the phasors of Figure 5.14 on the real axis, as the entire configuration rotates at the angular rate $\omega$. Further examples of phasor diagrams will be found in the problems at the end of the chapter.

## 5.3 POWER RELATIONSHIPS IN THE SINUSOIDAL STEADY STATE

### *Root-Mean-Square Value of a Sinusoid*

Let us suppose that a sinusoidal voltage is applied across the terminals of a resistor of $R$ ohms. For simplicity, let the voltage sinusoid be

$$v(t) = |V| \sin \omega t = V_m \sin \omega t, \qquad (5.61)$$

where we have designated $|V|$ as $V_m$ to emphasize that $|V|$ is the maximum or peak value of the sinusoid. The instantaneous power dissipated in the resistor is

$$p(t) = v(t)i(t) = \frac{v^2(t)}{R}$$

$$= \frac{V_m^2}{R} \sin^2 \omega t = \frac{V_m^2}{R}\left(\frac{1}{2} - \frac{1}{2}\cos 2\omega t\right) \qquad (5.62)$$

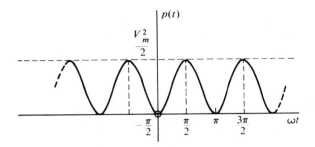

**Figure 5.15**  Instantaneous Power of Equation 5.62

and the waveform of this instantaneous power is sketched in Figure 5.15.

In most applications, the *time-average power* is a quantity of more interest than the instantaneous power. We may find the time average of $p(t)$ by integrating $p(t)$ over one period (or any integral number of periods), and dividing by the time interval. (In fact, we may find the time average of any periodic function in this way.) Denoting the time-average power by $P$, we have

$$P = \frac{V_m^2}{\pi R} \int_0^\pi \sin^2 x \, dx = \frac{V_m^2}{2R}. \tag{5.63}$$

We would like now to define an effective value of voltage, in such a way that the time-average power delivered by the sinusoidal voltage is equal to that which would be delivered if a constant voltage, $V_{\text{eff}}$, were applied across the resistor terminals. In other words, we want

$$\frac{V_{\text{eff}}^2}{R} = \frac{V_m}{2R}, \tag{5.64}$$

or

$$V_{\text{eff}} = \frac{V_m}{\sqrt{2}} = 0.707 V_m = V_{\text{RMS}}. \tag{5.65}$$

($V_{\text{RMS}}$ means *root-mean-square value* of voltage.) In similar fashion, we may define an RMS current as

$$I_{\text{RMS}} = I_{\text{eff}} = 0.707 I_m, \tag{5.66}$$

and thus we have

$$P = \frac{V_{\text{RMS}}^2}{R} = I_{\text{RMS}}^2 R = V_{\text{RMS}} I_{\text{RMS}}. \tag{5.67}$$

In practice, most sinusoidal voltages and currents are specified by giving their RMS values. (For example, 110 V RMS corresponds to a $V_m$ of 156 V.)

It should be noted that one may find the RMS value of any periodic wave-form $f(t)$, of period $T$, by integrating the square of $f(t)$ over a period, divid-ing by the period, and taking the square root of the result; i.e., for a periodic waveform of any shape,

$$\text{RMS value} = \left[ \frac{1}{T} \int_T f^2(t) \, dt \right]^{1/2}. \tag{5.68}$$

It should also be noted, however, that the particular relationship between the RMS and the "peak" or maximum value of a waveform as expressed by equations 5.65 and 5.66 is valid only for a sinusoid. The relationship for several other waveforms is illustrated in the problems at the end of the chapter.

### Power Factor and Complex Power

In the preceding section we calculated the time-average power supplied to a resistor by a sinusoidal voltage. In this case the voltage and current are in phase. Suppose now, however, that we have a case where the voltage and current are not in phase; for example, let

$$i(t) = |I| \sin \omega t \tag{5.69}$$

and

$$v(t) = |V| \sin (\omega t + \theta). \tag{5.70}$$

Now the time-average power will be

$$P = \frac{1}{\pi} \int_0^\pi v(t)i(t) \, d(\omega t)$$

$$= \frac{|V||I|}{\pi} \int_0^\pi \sin \omega t \sin (\omega t + \theta) \, d(\omega t)$$

$$= \frac{|V||I|}{\pi} \int_0^\pi \sin \omega t(\sin \omega t \cos \theta + \cos \omega t \sin \theta) \, d(\omega t)$$

$$= \frac{|V||I|}{\pi} \cos \theta \int_0^\pi \sin^2 \omega t \, d(\omega t) = \frac{|V||I|}{2} \cos \theta$$

$$= \frac{V_m I_m}{2} \cos \theta = V_{\text{RMS}} I_{\text{RMS}} \cos \theta. \tag{5.71}$$

Thus the fact that the voltage and current sinusoids are out of phase reduces the time-average power by the factor $\cos \theta$, where $\theta$, which is the relative phase angle, is called also *power-factor angle*. The factor $\cos \theta$ is called the *power factor*.

*Example 5.10*

For a pure resistor, $\theta = 0°$ and $\cos\theta = 1$. For a pure inductor or capacitor, $\theta = \pm 90$; and $\cos\theta = 0$. Thus no time-average power is delivered to a pure inductor or capacitor by a sinusoidal source.

*Example 5.11*

For the two circuits indicated in Figure 5.16, $\theta = \tan^{-1}(\omega L/R)$ and $-\tan^{-1}(I/\omega RC)$, respectively. In either case, as $R$ increases, the power factor approaches unity, and as $R$ decreases, the power factor approaches zero.

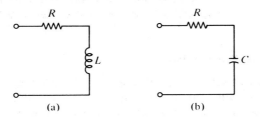

(a)                (b)

**Figure 5.16**  Circuits for Example 5.11

If we use the phasor model, our voltage and current are specified by a pair of complex numbers, $V$ and $I$, such that

$$v(t) = \text{Re}(V\epsilon^{j\omega t}) \tag{5.72}$$

$$i(t) = \text{Re}(I\epsilon^{j\omega t}) \tag{5.73}$$

and

$$|V| = V_m = \sqrt{2}\,V_{\text{RMS}} \tag{5.74}$$

$$|I| = I_m = \sqrt{2}\,I_{\text{RMS}}. \tag{5.75}$$

Then

$$\begin{aligned} P &= V_{\text{RMS}}I_{\text{RMS}}\cos\theta \\ &= \tfrac{1}{2}|V||I|\cos\theta \\ &= \tfrac{1}{2}[\text{Re}(V)\text{Re}(I) + I_m(V)I_m(I)] \\ &= \tfrac{1}{2}\text{Re}(VI^*), \end{aligned} \tag{5.76}$$

where $I^*$ denotes the complex conjugate of $I$. (The reader may wish to verify this last relationship for himself. Note that $P$ is just one half of the dot product between $V$ and $I$.) It is convenient, in light of equation 5.76, to define a *complex power* as

$$S \triangleq \tfrac{1}{2}VI^*. \tag{5.77}$$

The real part of $S$ is the real power,

$$P = \text{Re}\,S = |S|\cos\theta$$
$$= V_{\text{RMS}}I_{\text{RMS}}\cos\theta. \tag{5.78}$$

The imaginary part of $S$ is termed *reactive power*; i.e.,

$$Q \triangleq I_m S = |S|\sin\theta$$
$$= V_{\text{RMS}}I_{\text{RMS}}\sin\theta, \tag{5.79}$$

and this reactive power is associated with the energy storage in the reactive elements of a network, just as the real power is associated with the energy lost in the dissipative elements of the network. For a purely reactive network —i.e., one with no dissipative elements—the power is purely reactive, since $\theta = \pm 90°$ and $Q = V_{\text{RMS}}I_{\text{RMS}}$.

We may easily relate the complex power to the impedance and admittance of a network. For any pair of terminals we have defined the impedance as

$$Z(j\omega) = \frac{V}{I}. \tag{5.80}$$

Since $I = V/Z$ and $I^* = V^*/Z^*$, we have

$$S = \tfrac{1}{2}VI^* = \tfrac{1}{2}IZI^* = \tfrac{1}{2}|I|^2 Z$$
$$= I_{\text{RMS}}^2 Z, \tag{5.81}$$

or

$$S = \frac{1}{2}VI^* = \frac{1}{2}\frac{VV^*}{Z^*} = \frac{1}{2}\frac{|V|^2}{Z^*} = V_{\text{RMS}}^2 Y^*. \tag{5.82}$$

Thus, summarizing,

$$S = I_{\text{RMS}}^2 Z = V_{\text{RMS}}^2 Y^*. \tag{5.83}$$

Representing the impedance and admittance in terms of their real and reactive parts, resistance and reactance, and conductance and susceptance, respectively, i.e.,

$$Z = R + jX$$
$$Y = G + jB,$$

we have

$$P = I_{\text{RMS}}^2 R = V_{\text{RMS}}^2 G \tag{5.84}$$
$$Q = I_{\text{RMS}}^2 X = -V_{\text{RMS}}^2 B. \tag{5.85}$$

The unit *watt* has its usage restricted, so it applies only to real power—i.e., it is the unit of $P$, or $V_{RMS}I_{RMS} \cos \theta$. The unit of $Q$ is called the *VAR* (for volt-ampere reactive), and the unit of $V_{RMS}I_{RMS}$ is called the *volt-ampere*, or *VA*. In power applications, the usual units are *kilowatts* (kW), *kilovars* (kVAR), and *kilovolt-amperes* (kVA).

### Example 5.12

A current of $I_{RMS}$ amperes flows in the circuit of Figure 5.17. We will calculate the real and reactive power supplied by this current.

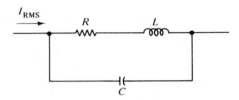

**Figure 5.17**   Circuit for Example 5.12

The impedance is

$$Z(j\omega) = \frac{(R + j\omega L)/j\omega C}{R + j\omega L + (1/j\omega C)}$$

$$= \frac{R + j\omega L}{(1 - \omega^2 LC) + j\omega RC} \frac{(1 - \omega^2 LC) - j\omega RC}{(1 - \omega^2 LC) - j\omega RC}$$

$$= \frac{[R(1 - \omega^2 LC) + \omega^2 RLC] + j[\omega L(1 - \omega^2 LC) - \omega R^2 C]}{(1 - \omega^2 LC)^2 + (\omega RC)^2}$$

$$= \frac{R + j\omega(L - \omega^2 L^2 C - R^2 C)}{(1 - \omega^2 LC)^2 + (\omega RC)^2} \tag{5.86}$$

Thus

$$P = \frac{RI_{RMS}^2}{(1 - \omega^2 LC)^2 + (\omega RC)^2}$$

and

$$Q = \frac{\omega I_{RMS}^2(L - \omega^2 L^2 C - R^2 C)}{(1 - \omega^2 LC)^2 + (\omega RC)^2}. \tag{5.87}$$

Note that as

$$\omega \longrightarrow 0, \qquad P \longrightarrow I_{RMS}^2 R,$$

and as

$$\omega \longrightarrow \infty, \qquad P \longrightarrow 0,$$

as would be expected from the high- and low-frequency equivalent circuits. At the resonant frequency, when $\omega^2 LC = 1$,

$$P = I_{RMS}^2 \frac{L}{RC}.$$

Will this be the frequency at which the power dissipated in the resistor is a maximum?

If an electrical load is connected to a source of constant RMS voltage, the RMS current must vary inversely as the cosine of the power factor angle if the real power is to remain constant. If the load is connected to the source by means of a transmission line with associated transformers, switchgear, etc., then not only must the source supply more current if the power factor angle is large, but all the transmission equipment must be able to handle the additional current. Hence power customers will often be penalized for poor power factors by rates that are adjusted accordingly, and power-factor correction (particularly the use of capacitors to correct for inductive loads which are typical of applications involving the use of heavy motors) often is desirable. An example is illustrated in a problem at the end of the chapter.

## 5.4 FREQUENCY RESPONSE AND SYSTEM FUNCTIONS

In Section 5.1 we saw that when a lumped linear network is driven by a sinusoidal forcing function, $x(t)$, then the response or output function $y(t)$ must also be sinusoidal, with the same frequency as the input. In phasor terms, if

$$x(t) = X\epsilon^{j\omega t}$$

and

$$y(t) = Y\epsilon^{j\omega t},$$

then (as in equations 5.18–5.20),

$$\frac{Y}{X} = H(j\omega) = \frac{N(j\omega)}{M(j\omega)}; \tag{5.88}$$

i.e., the $Y$ phasor is related to the $X$ phasor by a complex function of $\omega$, which takes the form of a rational fraction function of $(j\omega)$. $H(j\omega)$ is usually referred to as a *system function*; if $y(t)$ is a voltage and $x(t)$ a current, the system function $H(j\omega)$ is an impedance; if vice versa, it is an admittance. If these voltages and currents are defined to be those at a pair of input terminals, $H(j\omega)$ is an input or *driving-point impedance* or *admittance*. If they are defined at a pair of output terminals, $H(j\omega)$ is an *output impedance* or *admittance*. On the other hand, $x(t)$ might be the voltage at the input terminals, and $y(t)$ the voltage at the output terminals; in this case $H(j\omega)$ is usually called a *voltage transfer ratio*. We may also have, in similar fashion, a *current transfer ratio*. In short, we may consider any quantity in the network which we desire to be an output or response and relate it to any forcing-function quantity by the appropriate system function, as indicated in Figure 5.18.

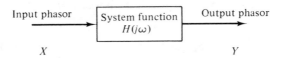

**Figure 5.18** Related Input and Output Phasors

Suppose that we now maintain a constant amplitude (say $|X| = 1$) on the input sinusoid but vary its frequency. For any frequency $\omega$ the output sinusoid will have an amplitude

$$|Y| = |H(j\omega)||X| = |H(j\omega)| \tag{5.89}$$

and a phase angle relative to the input given by

$$\underline{/Y} - \underline{/X} = \underline{/H(j\omega)}. \tag{5.90}$$

A specification or plot of $|H(j\omega)|$ as a function of $\omega$ is called an *amplitude response* of the circuit or system, and a specification or plot of $\underline{/H(j\omega)}$ is called a *phase response*. The two together constitute a *frequency response* for the system. Since $H(j\omega)$ is uniquely determined if its magnitude and angle are specified for all $\omega$, a complete frequency response is equivalent to the specification of the rational fraction of equation 5.88, from which the original differential equation relating input and output can be reconstructed using the operator notation of Section 5.1. Thus we arrive at the important conclusion that the input–output behavior of a circuit or system is completely determined by the frequency response—not only for sinusoidal inputs but for any inputs.

Methods of analysis and design based upon frequency responses play a very important role in electrical engineering, particularly in signal analysis, filter design, stability analysis, and control system analysis and design. The importance of frequency responses is further augmented by Fourier series and Fourier integral methods, which are discussed in Chapters 6 and 7.

### Polar Frequency-Response Plots

The magnitude and angle information contained in the system function may be displayed on a single diagram by the use of a polar plot, which simply shows the locus of $H(j\omega)$ in the complex plane as $\omega$ is varied from zero to infinity (or from minus infinity to plus infinity). This is illustrated by the following examples.

### Example 5.13

For the circuit of Figure 5.19, we will sketch a polar frequency response relating the input $v_1(t)$ and output $v_2(t)$. The appropriate system function is

$$H(j\omega) \triangleq \frac{V_2}{V_1} = \frac{1/j\omega C}{R + (1/j\omega C)} = \frac{1}{1 + j\omega RC}. \tag{5.91}$$

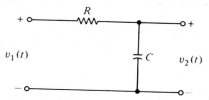

**Figure 5.19** Circuit for Example 5.13

Writing expressions for the magnitude and angle of $H(j\omega)$ we have

$$|H(j\omega)| = \frac{1}{\sqrt{1 + (\omega RC)^2}} \qquad (5.92)$$

$$\underline{/H(j\omega)} = -\tan^{-1} \omega RC. \qquad (5.93)$$

As $\omega$ varies from zero to infinity, $|H(j\omega)|$ goes from unity to zero, and the angle goes from zero to $-\pi/2$. In fact, a more detailed analysis (or an accurate plot) will show the locus to be a semicircular arc, as indicated in Figure 5.20.

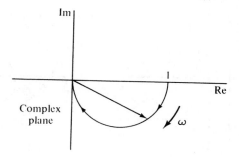

**Figure 5.20** Locus of $1/(1 + j\omega RC)$ in the Complex Plane

Although the polar plot shows clearly the magnitude and angle relationship of the frequency response, it has the drawback that values of $\omega$ corresponding to magnitudes and angles do not appear directly on an $\omega$ scale, as in the Bode plots discussed in the next section.

Several polar plots for typical system functions are shown in Example 5.14. The reader should verify the general shape of these plots by checking the magnitude and angle characteristics of the various system functions for low frequencies ($\omega \rightarrow 0$) and high frequencies ($\omega \rightarrow \infty$).

### Example 5.14

Polar frequency responses for some typical system functions are shown in Figure 5.21:

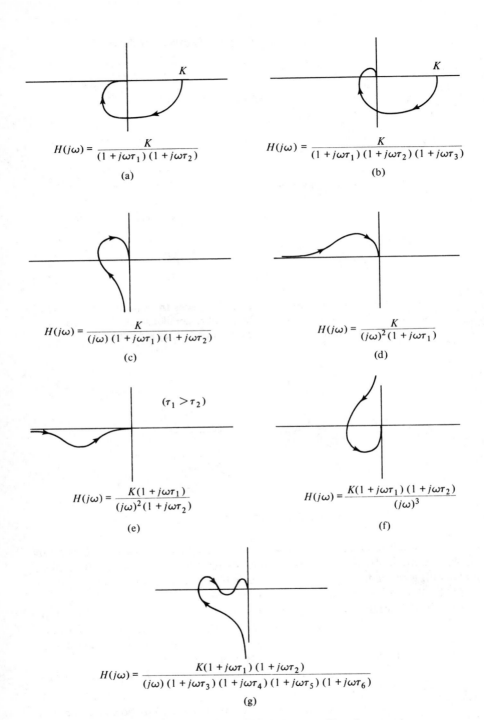

$$H(j\omega) = \frac{K}{(1 + j\omega\tau_1)(1 + j\omega\tau_2)}$$

(a)

$$H(j\omega) = \frac{K}{(1 + j\omega\tau_1)(1 + j\omega\tau_2)(1 + j\omega\tau_3)}$$

(b)

$$H(j\omega) = \frac{K}{(j\omega)(1 + j\omega\tau_1)(1 + j\omega\tau_2)}$$

(c)

$$H(j\omega) = \frac{K}{(j\omega)^2(1 + j\omega\tau_1)}$$

(d)

$$(\tau_1 > \tau_2)$$

$$H(j\omega) = \frac{K(1 + j\omega\tau_1)}{(j\omega)^2(1 + j\omega\tau_2)}$$

(e)

$$H(j\omega) = \frac{K(1 + j\omega\tau_1)(1 + j\omega\tau_2)}{(j\omega)^3}$$

(f)

$$H(j\omega) = \frac{K(1 + j\omega\tau_1)(1 + j\omega\tau_2)}{(j\omega)(1 + j\omega\tau_3)(1 + j\omega\tau_4)(1 + j\omega\tau_5)(1 + j\omega\tau_6)}$$

(g)

**Figure 5.21** Loci of Some Typical System Functions

172

Polar frequency-response plots such as those indicated in Figure 5.21 are essential in the application of *Nyquist's theorem*, which is a very important tool in the analysis of stability and performance of feedback control systems.

### Bode Frequency-Response Plots

Bode diagrams are plots of $|H(j\omega)|$ expressed in decibels, and $\underline{/H(j\omega)}$ in radians (or degrees), plotted against log $\omega$, or $\omega$ on a logarithmic scale. The decibel itself is a logarithmic unit, and $|H(j\omega)|$ in decibels is defined as

$$dB \triangleq 20 \log_{10} |H(j\omega)|. \tag{5.94}$$

(Hereafter, log will denote $\log_{10}$.) If, for example, $H(j\omega) = j\omega K$, as it would be if it represented the impedance of an inductor $(j\omega L)$ or the admittance of a capacitor $(j\omega C)$, then

$$|H(j\omega)| = K\omega \tag{5.95}$$

$$20 \log |H(j\omega)| = dB = 20 \log K + 20 \log \omega. \tag{5.96}$$

Thus, when plotted against log $\omega$, the dB curve is a straight line of slope $+20$ dB/decade[1] (or $+6$ dB/octave*, approximately), which intercepts the 0-dB axis at a frequency $\omega = 1/K$. Similarly, if $H(j\omega) = K/j\omega$, then

$$dB = 20 \log \frac{K}{\omega} = 20 \log K - 20 \log \omega, \tag{5.97}$$

and this is a straight line (as a function of log $\omega$) with a slope of $-20$ dB/decade, which intercepts the 0-dB axis at $\omega = K$. (Note that $\underline{/j\omega K} = \pi/2$, and $\underline{/K/j\omega} = -\pi/2$.)

Let us consider now the function

$$H(j\omega) = \frac{K}{1 + j\omega\tau}. \tag{5.98}$$

For example, if $K = 1$ and $\tau = RC$, this it the voltage transfer ratio of the circuit of Figure 5.19 (equation 5.91). Now, we may approximate $|H(j\omega)|$ as follows:

$$\omega < \frac{1}{\tau}, \quad |H(j\omega)| \simeq K, \quad dB \simeq 20 \log K \tag{5.99}$$

$$\omega > \frac{1}{\tau}, \quad |H(j\omega)| \simeq \frac{K}{\omega\tau}, \quad dB \simeq 20 \log \frac{K}{\tau} - 20 \log \omega. \tag{5.100}$$

---

[1]A *decade* is a frequency ratio of 10:1 and an *octave* is a frequency ratio of 2:1.

These two straight lines, the low- and high-frequency asymptotes, are shown in Figure 5.22(a). The two lines intersect at $\omega = 1/\tau$, which is sometimes referred to as the *break frequency*. This is also the frequency at which the straight-line approximation will be the most in error. In fact, at the frequency $\omega = 1/\tau$, the actual magnitude is

$$\left| H\left( j \frac{1}{\tau} \right) \right| = \frac{K}{\sqrt{2}}, \qquad (5.101)$$

or, in decibels,

$$\begin{aligned} \mathrm{dB} &= 20 \log K - 20 \log \sqrt{2} = 20 \log K - 10 \log 2 \\ &\simeq 20 \log K - 3. \end{aligned} \qquad (5.102)$$

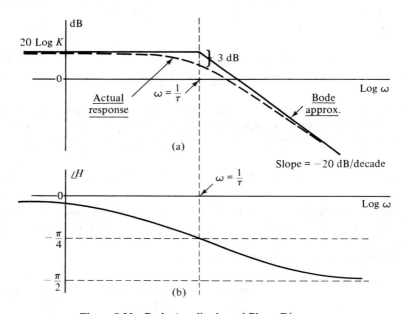

(a)

(b)

**Figure 5.22**  Bode Amplitude and Phase Diagram

Hence the response at the break frequency is about 3 dB down from the approximation. [The actual response is sketched as a dashed line on Figure 5.22(a).] For many analysis applications, the Bode or straight-line approximation is sufficiently accurate.

A sketch of the angle variation is shown in Figure 5.22(b). Note that at the break frequency the angle is $-\pi/4$; i.e.,

$$\left/ \frac{K}{1 + j1} \right. = 0 - \frac{\pi}{4} = -\frac{\pi}{4}. \qquad (5.103)$$

If we have the product of two or more system functions, the Bode amplitude plots simply add (which is one advantage of using a logarithmic scale), and the angle or phase responses also add. For example, if

$$H(j\omega) = \frac{1}{(1 + j\omega\tau_1)(1 + j\omega\tau_2)},$$  (5.104)

we have

$$dB = 20 \log \left| \frac{1}{1 + j\omega\tau_1} \right| + 20 \log \left| \frac{1}{1 + j\omega\tau_2} \right|$$  (5.105)

and

$$\underline{/H} = \underline{\left/ \frac{1}{1 + j\omega\tau_1} \right.} + \underline{\left/ \frac{1}{1 + j\omega\tau_2} \right.}.$$  (5.106)

Hence the Bode plot, for $\tau_1 > \tau_2$, would appear as in Figure 5.23. At each break frequency, the response is a little more than 3 dB down from the Bode approximation. (Note that if the two break frequencies coincided, the actual response would be 6 dB down, and the slope of the Bode approximation

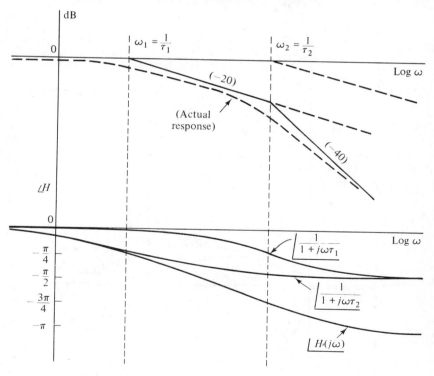

**Figure 5.23** Bode Plot Involving Two Factors

would go immediately from 0 dB/decade to $-40$ dB/decade.) The individual phase curves and their sum are also shown in the figure; the total phase angle excursion is from 0 to $-\pi$ radians.

Let us consider now the system function

$$H(j\omega) = \frac{K(1 + j\omega\tau_1)}{1 + j\omega\tau_2}, \qquad (5.107)$$

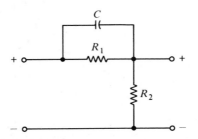

Figure 5.24   Phase-lead Network

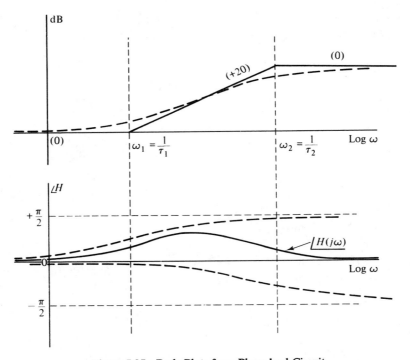

Figure 5.25   Bode Plots for a Phase-lead Circuit

with $\tau_1 > \tau_2$. This, for example, is the form of the voltage transfer ratio of the circuit of Figure 5.24, with

$$\frac{\tau_1}{\tau_2} = \frac{R_1 + R_2}{R_2}.$$

The numerator factor has a single break frequency, similar to the denominator factors considered previously, but the break is from a slope of 0 dB/decade to a slope of $+20$ dB/decade. When the denominator break occurs, at $\omega = 1/\tau_2$, the slope changes again by $-20$ dB/decade, leaving a net slope of zero for all higher frequencies. The Bode plots thus appear as in Figure 5.25. Since this circuit provides a positive or leading phase angle (i.e., the output sinusoid leads the input sinusoid), it is termed a *phase-lead network*, and has numerous applications, particularly in the design of feedback systems, where it is valuable as a stabilizing circuit.

If $H(j\omega)$ has the same form as in equation 5.107, but $\tau_1 < \tau_2$, the circuit is called a *phase-lag circuit*. The circuit realization and Bode diagram for a phase-lag circuit are left as an exercise (Problem 5.13).

### Example 5.15

A Bode diagram for the system function of Example 5.14(e) is sketched in Figure 5.26.

### Example 5.16

A bode diagram for the system function of Example 5.14(f) is sketched in Figure 5.27.

In all the examples of Bode diagrams examined thus far, the roots of both the numerator and denominator of $H(j\omega)$ have been real. If complex roots exist, we know that they must occur in conjugate pairs, and each conjugate pair yields a quadratic when the two conjugate complex factors are multiplied together. Thus, to be able to draw a Bode diagram for any rational $H(j\omega)$, we must investigate the Bode diagram for a quadratic with complex roots. Consider, for example, the system function

$$H(j\omega) = \frac{\omega_0^2}{(j\omega)^2 + 2\zeta\omega_0(j\omega) + \omega_0^2}. \tag{5.108}$$

A system function of this form results, for example, if we write the voltage transfer ratio of the circuit shown in Figure 5.28. This form of the quadratic factor, involving the damping ratio $\zeta$ and the undamped natural frequency $\omega_0$, has been used previously, in Chapter 4 (equation 4.137). For low frequencies, $|H(j\omega)| \simeq 1$, while for high frequencies $|H(j\omega)| \simeq 1/\omega^2$. Hence the low-frequency Bode asymptote is the 0-dB axis, while the high-frequency

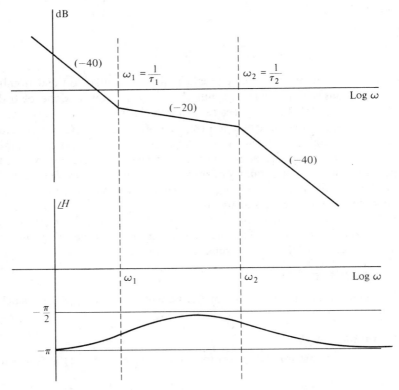

**Figure 5.26** Bode Diagram for $\dfrac{K(1 + j\omega\tau_1)}{(j\omega)^2(1 + j\omega\tau^2)}$

asymptote has a slope of $-40$ dB/decade. These two asymptotes intersect at $\omega = \omega_0$; in the vicinity of $\omega_0$, however, the actual frequency response may deviate greatly from these asymptotes, depending on the value of $\zeta$. A sketch of the Bode diagram for several values of $\zeta$ is shown in Figure 5.29. More accurate Bode plots of a quadratic term may be found in almost any textbook on control theory. With such a plot available, and with the knowledge of the Bode behavior of factors involving real roots, the Bode diagram for any rational $H(j\omega)$ can be readily obtained (assuming, of course, that one is able to factor the numerator and denominator polynomials). The key to the ease of obtaining the complete Bode diagram is the fact that both the dB and angle plots for the various individual factors of $H(j\omega)$ are additive.

Both polar and Bode frequency-response plots are important tools in analysis and design, particularly when one is dealing with circuits involving feedback, and feedback control systems.

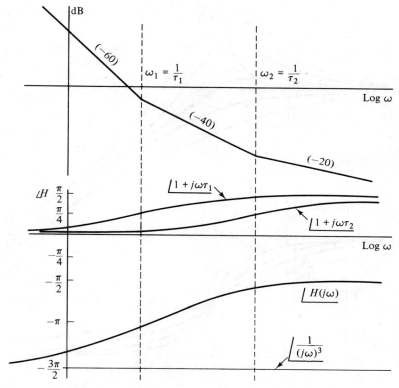

**Figure 5.27** Bode Diagram for $\dfrac{K(1 + j\omega\tau_1)(1 + j\omega\tau_2)}{(j\omega)^3}$

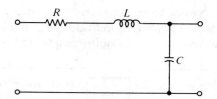

**Figure 5.28** Circuit Having the Voltage Transfer
Ratio of Equation 5.108

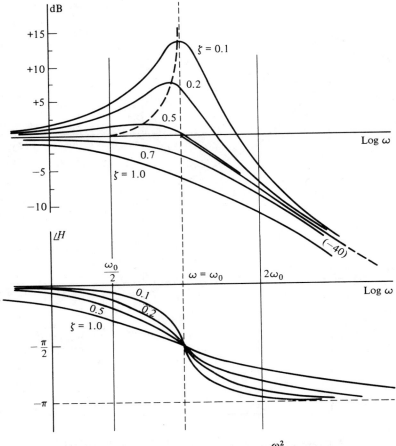

**Figure 5.29** Bode Diagram for $\dfrac{\omega_0^2}{(j\omega)^2 + 2\zeta\omega_0(j\omega) + \omega_0^2}$

## 5.5 THÉVENIN'S AND NORTON'S THEOREMS

In most applications of electrical networks, it is convenient to consider a part of the total network to be the *source* or *driving* part of the network, and the remainder to be the *load*, as indicated Figure 5.30. Both the driving and load parts of the network may be quite complicated networks themselves,

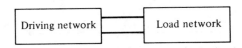

**Figure 5.30** Driving and Load Parts of a Network

but we will assume that they are linear, and, for our sinusoidal steady-state analysis, time-invariant. (Later we shall see that these theorems also apply in networks where the forcing functions are not sinusoidal; i.e., their validity is not restricted to networks operating in the sinusoidal steady state.) The theorems state, in effect, that the driving part of the network may be replaced by an equivalent circuit consisting of a single independent source and a single impedance, insofar as its terminal behavior, or its effect on the load network, is concerned. Detailed proofs of these theorems are somewhat lengthy and will be omitted here; rigorous proofs may be found in many texts on circuit theory.[2]

### Thévenin's Theorem

Thévenin's theorem may be expressed by the following three statements:

1. Any network, insofar as its terminal characteristics are concerned, may be replaced by an equivalent circuit consisting of a single independent voltage source in series with a single impedance.
2. The value of the Thévenin equivalent voltage source is the voltage that appears at the open-circuited terminals of the original network.
3. The value of the Thévenin equivalent impedance is the impedance seen looking back into the circuit from its open-circuited terminals, with all independent sources replaced by their internal impedances (i.e., all independent ideal voltage sources short-circuited and all independent ideal current sources replaced by open circuits—in other words, all independent sources reduced to zero value).

Thus the Thévenin equivalent network takes the form of the circuit of Figure 5.31. Note that, for this circuit, the open-circuited terminal voltage is just $V_T$, and the impedance seen from the terminals with the voltage source replaced by a short circuit is just $Z_T$. Although we will not give a proof here of this theorem, we will illustrate its application by means of examples.

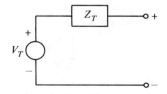

**Figure 5.31** Thévenin Equivalent Circuit

---

[2]See, for example, P. M. Chirlian, *Basic Network Theory*, McGraw-Hill Book Company, New York, 1969.

**Example 5.17**

We will find the Thévenin equivalent of the circuit shown in Figure 5.32. If we look back from the output terminals with the voltage source replaced by a short circuit and the current source replaced by an open circuit, we see the Thévenin equivalent impedance

$$Z_T = \frac{R_1 R_2}{R_1 + R_2}. \qquad (5.109)$$

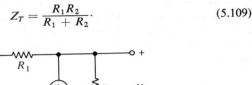

**Figure 5.32**   Circuit for Example 5.17

To find the Thévenin equivalent voltage source, $V_T$, we may use Kirchhoff's current law and write

$$\frac{V - V_T}{R_1} + I - \frac{V_T}{R_2} = 0, \qquad (5.110)$$

or, solving for $V_T$,

$$V_T = Z_T \left( \frac{V}{R_1} + I \right). \qquad (5.111)$$

These then are the quantities that constitute the equivalent circuit of Figure 5.31.

**Example 5.18**

We wish to break the circuit of Figure 5.33 at a point indicated by the dashed line and replace everything to its left by the Thévenin equivalent. If we break the circuit at the dashed line and look back into the terminals

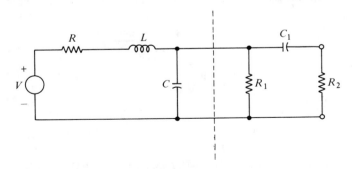

**Figure 5.33**   Circuit of Example 5.18

thus formed, with the voltage source replaced by a short circuit, we see an impedance

$$Z_T = \frac{(1/j\omega C)(R + j\omega L)}{(1/j\omega C) + R + j\omega L} = \frac{R + j\omega L}{(1 - \omega^2 LC) + j\omega RC}. \qquad (5.112)$$

The voltage that appears at the open-circuited terminals is (using the voltage divider formula)

$$V_T = \frac{(1/j\omega C)V}{R + j\omega L + (1/j\omega C)} = \frac{V}{(1 - \omega^2 LC) + j\omega RC}.$$

Thus the circuit of Figure 5.33 may be replaced by that of Figure 5.34.

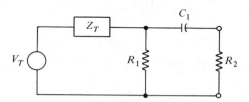

**Figure 5.34**  Thévenin Equivalent for Figure 5.33

### *Example 5.19*

If a dependent source appears in the original circuit, that source is *not* reduced to zero value when calculating the Thévenin equivalent impedance. Let us, for example, find the Thévenin equivalent of the circuit of Figure 5.35, which involves an independent voltage source and a dependent current source.

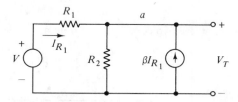

**Figure 5.35**  Circuit of Example 5.19

To evaluate $V_T$, we may apply Kirchhoff's current law. First we write the current equation at node $a$:

$$I_{R_1} + \beta I_{R_1} + (1 + \beta)I_{R_1} = \frac{V_T}{R_2}. \qquad (5.113)$$

The current through the resistor $R_1$ is given by

$$I_{R_1} = \frac{V - V_T}{R_1}. \qquad (5.114)$$

If we combine the foregoing two equations and solve for $V_T$, we obtain

$$\frac{V}{R_1} = V_T \left[ \frac{1}{R_1} + \frac{1}{(1 + \beta)R_2} \right] \tag{5.115}$$

$$V_T = \frac{(1 + \beta)R}{R_1 + (1 + \beta)R_2} V. \tag{5.116}$$

To find $Z_T$ we replace the independent voltage source by a short circuit and calculate the impedance as seen from the terminals; this will be the ratio of $V_0$ to $I_0$, as indicated in Figure 5.36. Again, by application of Kirchhoff's current law at node $a$:

$$(1 + \beta)I_{R_1} + I_0 = \frac{V_0}{R_2}. \tag{5.117}$$

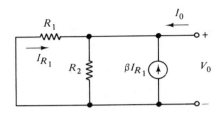

**Figure 5.36** Circuit for Calculating $Z_T$ in Example 5.19

The voltage across $R_1$ is now just $(-V_0)$; hence

$$I_{R_1} = -\frac{V_0}{R_1}. \tag{5.118}$$

If we combine these two equations, we obtain

$$I_0 = V_0 \left[ \frac{1}{R_2} + \frac{(1 + \beta)}{R_2} \right]$$

$$Z_T = \frac{V_0}{I_0} = \frac{R_1 R_2}{R_1 + (1 + \beta)R_2}. \tag{5.119}$$

As indicated by this example, it is generally necessary to go back to the circuit equations as derived from Kirchhoff's laws to find the Thévenin equivalent of a network that contains dependent sources.

### Norton's Theorem

If we do a source transformation of the kind discussed in Chapter 3 (Figure 3.30), our Thévenin equivalent circuit may be converted into another equivalent circuit involving a current source and a shunt impedance, as indicated in Figure 5.37(b). If we apply Thévenin's theorem to the circuit of part (b) of the figure, it is clear that

$$Z_N = Z_T \qquad (5.120)$$

and

$$I_N = \frac{V_T}{Z_T}. \qquad (5.121)$$

Thus, to find the Norton equivalent impedance, for any network, we proceed exactly as in finding the Thévenin equivalent impedance. To find the Norton equivalent current source, we may use equation 5.121, or, referring to Figure 5.37(b), we may short-circuit the terminals of our network and calculate the current which flows in that short circuit. Examples appear in the problems at the end of the chapter.

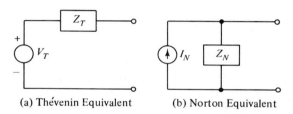

(a) Thévenin Equivalent          (b) Norton Equivalent

**Figure 5.37**   Thévenin and Norton Equivalent Circuits

### 5.6   SUPERPOSITION

Thévenin's and Norton's theorems provide useful tools in network analysis in that they allow the reduction of complicated networks to networks of simpler form. Superposition is another tool, which is especially valuable in analyzing networks containing more than one independent source. In Section 3.4 we discussed the concept and implications of linearity; the use of superposition is simply an application of linearity. If we wish to evaluate some quantity in a linear network which results from the application of more than one source, we may first evaluate the quantity due to each individual source, with the others reduced to zero value, and then sum the results. (Again, to reduce an ideal voltage source to zero value, we replace it by a short circuit; and to reduce an ideal current source to zero value, we replace it by an open circuit.) The use of superposition is illustrated by the following example.

#### *Example 5.20*

We will use superposition to evaluate $V_T$ in the circuit of Figure 5.32.

First, let us open-circuit the current source and evaluate the $V_T$ that results from the voltage source. Using the voltage-divider formula, we get

$$V_{T_1} = \frac{VR_2}{R_1 + R_2}.$$  (5.122)

Next, we short-circuit the voltage source and evaluate the $V_T$ which results from the current source (with the voltage source shorted, $R_1$ and $R_2$ appear in parallel):

$$V_{T_2} = \frac{IR_1R_2}{R_1 + R_2}.$$  (5.123)

Then

$$V_T = V_{T_1} + V_{T_2} = \frac{R_2(V + R_1I)}{R_1 + R_2},$$  (5.124)

which is identical to the expression (equation 5.111) previously obtained using Thévenin's theorem.

## PROBLEMS

**5.1** In each of the following equations, $f(t)$ is a sinusoidal forcing function. Find the ratio of the peak values of $x(t)$ and $f(t)$ for each equation, and the phase difference between $x(t)$ and $f(t)$.
(a) $\dot{x}(t) + ax(t) = bf(t)$.
(b) $\dddot{x}(t) + \ddot{x}(t) + \dot{x}(t) + x(t) = f(t)$.
(c) $\ddot{x}(t) + ax(t) = f(t) + bf(t)$.
(d) $\ddot{x}(t) + 2\dot{x}(t) + 3x(t) = f(t) + 4\dot{f}(t)$.

**5.2** For the equations of Problem 5.1, write an expression for $x(t)$ if
(a) $f(t) = \sin t$.
(b) $f(t) = 2\cos(3t + 1)$.

**5.3** Find the impedance, admittance, reactance, and susceptance for the circuit of Figure 5.38.

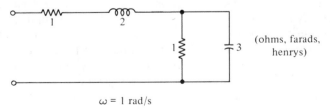

$\omega = 1$ rad/s

**Figure 5.38** Circuit for Problem 5.3

**5.4** Sketch a phasor diagram, approximately to scale, showing every voltage, current, charge, and flux linkage phasor in the circuit of Figure 5.38.

**5.5** (a) Find the voltage transfer ratio $V_2(j\omega)/V_1(j\omega)$, for the circuit of Figure 5.39.

(b) If $v_1(t) = \cos t$ in Figure 5.39, what is $v_2(t)$?

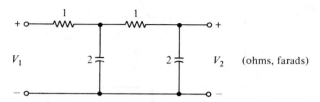

**Figure 5.39** Circuit of Problem 5.5

**5.6** The circuit shown in Figure 5.40 is operating in the sinusoidal steady state. Sketch a phasor diagram, approximately to scale, showing phasors for each voltage and current in the circuit. (*Hint:* Start with the current in the 2-Ω resistor.)

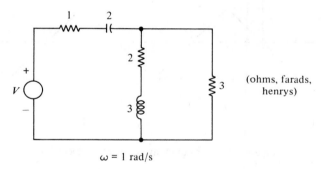

$\omega = 1$ rad/s

**Figure 5.40** Circuit for Problem 5.6

**5.7** In the transistor circuit of Figure 3.34, $v(t) = \cos t$, and all element values are unity. Find the current in the load resistor $R$, the RMS value of this current, and the average power dissipated in the load.

**5.8** Find the average values and the RMS values of the periodic waveforms sketched in Figure 5.41. Sketch the instantaneous power in each case.

**5.9** The circuit of Figure 5.38 is connected to a sinusoidal voltage source of 5 V RMS, and a frequency of 10 cycles per second (hertz). Find the complex power supplied by the source, the real power dissipated in the circuit, and the power factor.

**5.10** In the inductive curcuit of Figure 5.42, it is desired to correct the power factor by inserting a capacitor into the circuit in series with the 1-Ω resistor. What value of capacitor will be required to make the power factor unity?

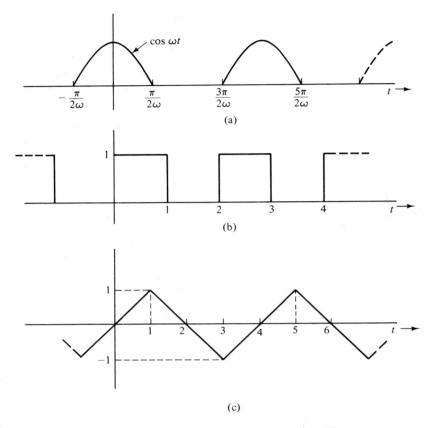

**Figure 5.41**   Periodic Waveforms of Problem 5.8

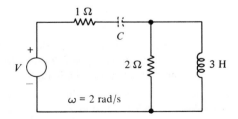

**Figure 5.42**   Circuit of Problem 5.10

**5.11**   An approximate equivalent circuit for a transformer is shown in Figure 5.43. If the operating frequency is 60 Hz, $L_1 = L_2 = 10$ H, and $M = 9$ H, find the voltage transfer ratio of the transformer.

**5.12**   Sketch a Bode diagram for the frequency response of the circuit of Figure 5.43 if $R_L = 2\ \Omega$.

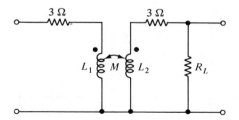

**Figure 5.43**   Circuit of Problem 5.11

**5.13**   Show a circuit realization (using only resistors and capacitors) for a phase-lag network (equation 5.107, with $\tau_1 < \tau_2$). Find expressions for $\tau_1$ and $\tau_2$ in terms of the element values, and sketch a Bode frequency-response diagram (magnitude and phase).

**5.14**   Sketch a polar frequency-response plot for the phase-lag network of Problem 5.13. Compare with the polar plot for a phase-lead network.

**5.15**   Sketch a Bode approximation for the magnitude and angle of the following system functions:

(a) $H(j\omega) = \dfrac{100(j0.1\omega + 1)}{(j\omega + 1)(j0.01\omega + 1)(j0.001\omega + 1)}$.

(b) $H(j\omega) = \dfrac{1}{j\omega(j\omega + 1)^2}$.

(c) $H(j\omega) = \dfrac{1}{(j\omega)^2(j\omega + 1)}$.

(d) $H(j\omega) = \dfrac{(j\omega + 1)^2}{(j\omega)^3}$.

**5.16**   Sketch polar frequency-response plots for the system functions of Problem 5.15.

**5.17**   A pure time delay is represented by the system function

$$H(j\omega) = \epsilon^{-j\omega T},$$

where $T$ is the delay time in seconds. Sketch both a Bode and a polar frequency response for this system function. From the frequency response, can you explain why an input sinusoid of any frequency is simply delayed (shifted in time) by $T$ seconds?

**5.18**   Find both Thévenin and Norton equivalents for the circuits shown in Figure 5.44.

**5.19**   The circuit of Figure 5.43 is driven by a sinusoidal voltage source with a frequency of 60 Hz. Find a Thévenin equivalent if the circuit is broken at the terminals of the load resistor, $R_L$.

**5.20**   Use superposition to find the terminal voltage of the circuit of Figure 5.44(b).

**5.21**   Use superposition to find the current in the load resistor $R_L$ in the circuit of Figure 5.45.

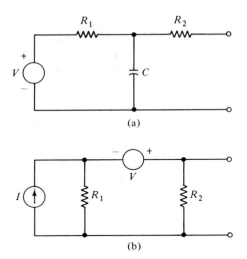

(a)

(b)

**Figure 5.44** Circuits of Problem 5.18

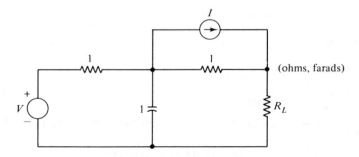

**Figure 5.45** Circuit of Problem 5.21

**5.22** For a network having the system function of equation 5.107, show that
(a) $\underline{/H(j\omega_1)} = \underline{/H(j\omega_2)}$, where $\omega_1 = 1/\tau_1$ and $\omega_2 = 1/\tau_2$.
(b) $\underline{/H(j\omega)}$ reaches a maximum (or minimum) at the midpoint between $\omega_1$ and $\omega_2$, on a logarithmic frequency scale.

**5.23** In a second-order system, show that the expression for $f(t)$ in the case of critical damping can be obtained by starting with expression in the under-damped case and considering the limit as $\zeta$ approaches unity.

# 6

# *Fourier Series and Fourier Transforms*

At the beginning of Chapter 5 it was stated that one reason for the importance of sinusoidal-steady-state analysis is that it may be extended to the analysis of circuits or systems involving other kinds of signals—i.e,, nonsinusoidal signals. In particular, any periodic signal (with certain restrictions that are usually not important for physical systems) may be expressed as a sum of sinusoids. By a limiting process, this sum may take the form of an integral that is useful in representing nonperiodic signals as well. In this chapter we shall elaborate on the basic idea of representing nonsinusoidal signals as sums of sinusoids.

### 6.1  PERIODIC FUNCTIONS AND THE DIRICHLET CONDITIONS

Any function that repeats itself every $T$ seconds is said to be periodic, with a period $T$. Mathematically, any function such that

$$f(t + T) = f(t) \qquad (-\infty < t < +\infty) \qquad (6.1)$$

is periodic, and has a frequency, in cycles per second (Hz), given by

$$f = \frac{1}{T}. \qquad (6.2)$$

Typical periodic functions of great engineering importance include sinusoids, square waves, triangular waves, pulse trains, and a multitude of variously shaped periodic waves generated by different types of nonlinear oscillators.

In our discussion of periodic functions, we will restrict ourselves to functions that belong to a certain class—functions that satisfy the *Dirichlet conditions*. A function belongs to this class if it is everywhere finite, has at most a finite number of maxima and minima in any period, has at most a finite number of discontinuities in any period,[1] and is absolutely integrable over any period, i.e., if

$$\int_T |f(t)|\, dt < \infty. \tag{6.3}$$

[This notation indicates that the integral may be taken over any period, such as $(-T/2, +T/2)$, $(0, T)$, $(T, 2T)$, etc.; all the integrals will be the same.] Most periodic signals of engineering interest, such as those enumerated earlier, satisfy these conditions, but many periodic mathematical functions do not.

### Example 6.1

The periodic function $f(t) = \tan(t)$ does not satisfy the Dirichlet conditions. Why?

### Example 6.2

A periodic function may always be generated by taking a piece of a non-periodic function and simply repeating it every $T$ seconds. If we start with the function

$$f(t) = \sin\left(\frac{1}{t}\right) \qquad \left(-\frac{T}{2} < t < \frac{T}{2}\right)$$

and repeat it every $T$ seconds, the resulting periodic function does not satisfy the Dirichlet conditions. Why?

### 6.2 SINE–COSINE FORM OF THE FOURIER SERIES

The Fourier theorem states that any periodic function which satisfies the Dirichlet conditions can be expanded in an infinite series of the form

$$f(t) = a_0 + \sum_{k=1}^{\infty} (a_k \cos k\omega_0 t + b_k \sin k\omega_0 t), \tag{6.4}$$

where $\omega_0 = 2\pi/T$. If we examine this series, we note that $a_0$ is simply a constant component, which is the average value of the periodic function (since the sinusoidal components all have zero average values). The lowest frequency

---

[1] A function having these three properties is also said to be a function of *bounded variation*.

sinusoids of the series are associated with $k = 1$, so these have a frequency of $\omega_0$ radians/second, which is called the *fundamental radian frequency*. All the other sinusoids in the series have frequencies that are integer multiples of $\omega_0$; these are termed *harmonics*.

To find the Fourier series expansion of some given periodic function, we must evaluate all of the $a$ and $b$ coefficients. This can be done by making use of certain properties of sinusoids called *orthogonality properties*. Two functions, $f_1(t)$ and $f_2(t)$, are said to be *orthogonal* over an interval $(t_1, t_2)$ if

$$\int_{t_1}^{t_2} f_1(t) f_2(t) \, dt = 0. \tag{6.5}$$

For sinusoids we have the following relationships (with $\omega_0 = 2\pi/T$):

$$\int_{-T/2}^{+T/2} \sin (m\omega_0 t) \sin (n\omega_0 t) \, dt = \begin{cases} 0 & m \neq n \\ \dfrac{T}{2} & m = n \end{cases}$$

$$\int_{-T/2}^{+T/2} \cos (m\omega_0 t) \cos (n\omega_0 t) \, dt = \begin{cases} 0 & m \neq n \\ \dfrac{T}{2} & m = n \end{cases}$$

$$\int_{-T/2}^{+T/2} \sin (m\omega_0 t) \cos (n\omega_0 t) \, dt = 0 \qquad \text{(all } m \text{ and } n\text{).} \tag{6.6}$$

Thus to evaluate $a_n$, for $n \neq 0$, we need simply to multiply both sides of equation 6.4 by $\cos (n\omega_0 t)$, and integrate over one period, as indicated in equation 6.7.

$$\int_{-T/2}^{+T/2} f(t) \cos n\omega_0 t \, dt$$

$$= \int_{-T/2}^{+T/2} \cos n\omega_0 t \left[ a_0 + \sum_{k=1}^{\infty} (a_k \cos k\omega_0 t + b_k \sin k\omega_0 t) \, dt. \right. \tag{6.7}$$

If we integrate the right-hand side of equation 6.7 term by term (which we assume can be done) and make use of the orthogonality relationships of equations 6.6, all terms are zero except that involving $a_n$, and we thus obtain

$$a_n = \frac{2}{T} \int_{-T/2}^{+T/2} f(t) \cos n\omega_0 t \, dt. \tag{6.8}$$

Similarly,

$$b_n = \frac{2}{T} \int_{-T/2}^{+T/2} f(t) \sin n\omega_0 t \, dt. \tag{6.9}$$

For $n = 0$ we have only the $a_0$ coefficient, and since this is simply the average value of $f(t)$,

$$a_0 = \frac{1}{T} \int_{-T/2}^{+T/2} f(t)\, dt. \tag{6.10}$$

These formulas allow us to evaluate all the coefficients of the Fourier series. It should be noted that the integrals may be taken over any interval, such as $(-T/2, T/2)$ or $(0, T)$, provided that the interval is of one period (or any integer multiple of one period) duration.

### Example 6.3

We will find the Fourier series expansion for the square wave of Figure 6.1. For this function, $T = 2$, $\omega_0 = \pi$, and $a_0 = 0$. If we choose $(0, 2)$ as our interval (of one period duration) over which to integrate, we get

$$
\begin{aligned}
a_n &= \tfrac{2}{2} \int_0^2 f(t) \cos n\pi t\, dt \\
&= \int_0^1 (1) \cos n\pi t\, dt + \int_1^2 (-1) \cos n\pi t\, dt \\
&= 0.
\end{aligned}
\tag{6.11}
$$

$$
\begin{aligned}
b_n &= \tfrac{2}{2} \int_0^2 f(t) \sin n\pi t\, dt \\
&= \int_0^1 (1) \sin n\pi t\, dt + \int_1^2 (-1) \sin n\pi t\, dt \\
&= \frac{2}{n\pi}(1 - \cos n\pi) = \begin{cases} \dfrac{4}{n\pi} & (n = 1, 3, 5, \ldots) \\ 0 & (n = 2, 4, 6, \ldots). \end{cases}
\end{aligned}
\tag{6.12}
$$

Thus the Fourier series expansion for the square wave of Figure 6.1 is

$$
\begin{aligned}
f(t) &= \frac{4}{\pi}\left( \sin \pi t + \frac{1}{3} \sin 3\pi t + \frac{1}{5} \sin 5\pi t + \ldots \right) \\
&= \frac{4}{\pi} \sum_{k=1,3,5,\ldots}^{\infty} \frac{1}{k} \sin k\pi t.
\end{aligned}
\tag{6.13}
$$

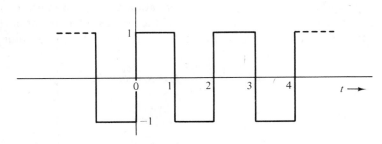

**Figure 6.1**  Square Wave of Example 6.3

### Example 6.4

We will find the Fourier series expansion of the pulse train shown in Figure 6.2. Each pulse has a height $H$, a duration $w$ seconds, and a repetition period of $T$ seconds. We will take our integrals over the interval $(-T/2, +T/2)$.

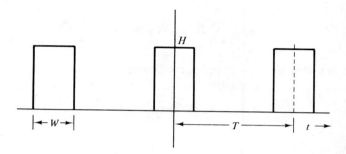

**Figure 6.2** Pulse Train of Example 6.4

$$a_0 = \frac{1}{T} \int_{-T/2}^{+T/2} f(t)\, dt = \frac{H}{T} \int_{-w/2}^{+w/2} dt = \frac{wH}{T}. \qquad (6.14)$$

$$a_n = \frac{2}{T} \int_{-T/2}^{+T/2} f(t) \cos n\frac{2\pi}{T} t\, dt$$

$$= \frac{2H}{T} \int_{-w/2}^{+w/2} \cos n\frac{2\pi}{T} t\, dt$$

$$= \frac{H}{n\pi} \sin n\frac{2\pi}{T} t \Big|_{-w/2}^{+w/2} = \frac{2H}{n\pi} \sin \frac{n\pi w}{T}. \qquad (6.15)$$

$$b_n = \frac{2H}{T} \int_{-w/2}^{+w/2} \sin n\frac{2\pi}{T} t\, dt = 0. \qquad (6.16)$$

Hence the Fourier series expansion of the pulse train is

$$f(t) = \frac{wH}{T} + \frac{2H}{\pi} \sum_{n=1}^{\infty} \frac{1}{n} \sin \frac{n\pi w}{T} \cos n\frac{2\pi}{T} t. \qquad (6.17)$$

## 6.3 EXPONENTIAL FORM OF THE FOURIER SERIES

A more concise and theoretically useful form of the Fourier series may be arrived at by making use of Euler's relationships (equations 4.37). Substitution of

$$\cos k\omega_0 t = \frac{1}{2}(\epsilon^{jk\omega_0 t} - \epsilon^{-jk\omega_0 t})$$

and

$$\sin k\omega_0 t = \frac{1}{2j}(\epsilon^{jk\omega_0 t} - \epsilon^{-jk\omega_0 t})$$

into equation 6.4 yields

$$f(t) = a_0 + \sum_{k=1}^{\infty} \frac{a_k - jb_k}{2} \epsilon^{jk\omega_0 t} + \frac{a_k + jb_k}{2} \epsilon^{-jk\omega_0 t}. \qquad (6.18)$$

If we now define a new complex coefficient,

$$\alpha_k \triangleq \frac{a_k - jb_k}{2} \qquad (6.19)$$

and

$$\alpha_{-k} \triangleq \alpha_k^* = \frac{a_k + jb_k}{2}, \qquad (6.20)$$

the entire series of equation 6.18 may be written as

$$f(t) = \sum_{k=-\infty}^{+\infty} \alpha_k \epsilon^{jk\omega_0 t}. \qquad (6.21)$$

Note that this expression produces the $a_0$ term of equation 6.18 (i.e., $\alpha_0 = a_0$, the $k = 0$ term of the series), and also produces all the negative exponentials since the range of $n$ has been extended to include all negative integers as well as the positive integers. Equation 6.21 is called the *exponential form of the Fourier series* and is simply an alternative way of writing equation 6.4.

We may evaluate the complex coefficients $\alpha_k$ either by calculating $a_k$ and $b_k$ as before and using equation 6.19, or we may calculate them directly by integration. Rewriting the integrals for $a_k$ and $b_k$ in exponential form by using Euler's relationships and substituting them into equation 6.19 (see Problem 6.6), we obtain

$$\alpha_k = \frac{1}{T} \int_{-T/2}^{+T/2} f(t)\epsilon^{-jk\omega_0 t} \, dt. \qquad (6.22)$$

One advantage of this form of the Fourier series is that only one formula (equation 6.22) is required to express all the coefficients, including $\alpha_0$, whereas previously three formulas were required (for $a_0$, $a_k$, and $b_k$). We may note also, from equation 6.19, that $\alpha_k$ will be real for any series involving only cosine terms, imaginary for any series involving only sine terms, and complex for a series with both sine and cosine terms.

### Example 6.5

Let us return to Example 6.3, and express the Fourier series of the square wave of Figure 6.1 in complex form. Since we have already found $a_n$ and $b_n$, we may use equation 6.19 to find $\alpha_n$:

$$\alpha_n = \frac{a_n - jb_n}{2} = \begin{cases} -\dfrac{j2}{n\pi} & (n = \pm 1, \pm 3, \pm 5, \ldots) \\ 0 & (n = 0, \pm 2, \pm 4, \ldots) \end{cases} \tag{6.23}$$

Thus the series expansion may be written

$$f(t) = -\frac{j2}{\pi} \sum_{n=-\infty}^{+\infty} \frac{1}{n} \epsilon^{jn\pi t}. \tag{6.24}$$
$$(n \text{ odd})$$

The same coefficients will be obtained if we evaluate them using equation 6.22. Note that these coefficients are all imaginary, because the waveform expansion contains only sine terms.

### Line Spectra

Let us investigate the relationship between the complex coefficient $\alpha_n$ and the amplitude and phase angle of the corresponding sinusoidal component of the Fourier series. Each coefficient $\alpha_n$ is associated with a sinusoidal component

$$s(t) = \alpha_n \epsilon^{jn\omega_0 t} + \alpha_n^* \epsilon^{-jn\omega_0 t}. \tag{6.25}$$

(Here we have grouped together the corresponding positive and negative index terms of the exponential series.) We know, however, that equation 6.25 may be written in the form (see Chapter 4)

$$s(t) = 2|\alpha_n| \cos (n\omega_0 t + \underline{/\alpha_n}). \tag{6.26}$$

Thus the amplitude of each sinusoidal component is twice the magnitude of $\alpha_n$, and its phase angle is the angle or argument of $\alpha_n$. In other words, $\alpha_n$ is simply a phasor whose length is half the amplitude of the sinusoidal component. If we "plot" $2|\alpha_n|$ and $\underline{/\alpha_n}$ for each positive $n$, we have what is often referred to as a *single-sided* amplitude and phase spectrum, respectively. Alternatively, we might plot $|\alpha_n|$ and $\underline{/\alpha_n}$ for both positive and negative $n$; these are often called *double-sided* amplitude and phase spectra. In either case, the interpretation of these line spectra in terms of the amplitudes and phase angles of the component sinusoids is clear from equation 6.26. Note that, since $\alpha_{-n} = \alpha_n^*$, $|\alpha_{-n}| = |\alpha_n|$ and $\underline{/\alpha_{-n}} = -\underline{/\alpha_n}$, as illustrated in Figure 6.3.

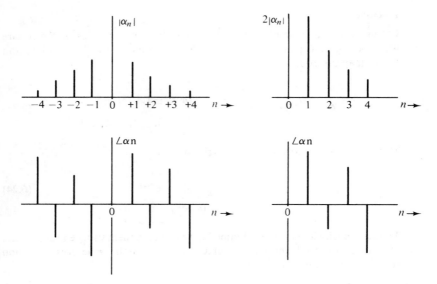

**Figure 6.3**  Double- and Single-sided Line Amplitude and Phase Spectra

Let us now look at the *time-average power* contributed by each sinusoidal component. We know, from Chapter 5, that the time-average power associated with a signal is proportional to its mean-square value, or the square of the RMS value, which is given by

$$P_n = S_{\text{RMS}}^2 = \frac{S_{\text{max}}^2}{2} = \frac{4\,|\alpha_n|^2}{2} = 2\,|\alpha_n|^2.$$

Thus, if we plot $|\alpha_n|^2$ for all values of $n$, positive and negative, we have a double-sided line spectrum, but each line now tells us how much time-average power is associated with each value of $n$, and it may be shown (see Problem 6.9) that the total time-average power of the original waveform is obtained by simply adding up the powers associated with all the spectral lines; i.e.,

$$P_{\text{tot}} = \sum_{n=-\infty}^{\infty} |\alpha_n|^2. \tag{6.27}$$

Similarly, it may be shown that within a given bandwidth, say $-N \leq n \leq N$, the total power contribution within that bandwidth is given by

$$P_{\text{partial}} = \sum_{n=-N}^{+N} |\alpha_n|^2. \tag{6.28}$$

The plot of $|\alpha_n|^2$ versus $n$ is called a *discrete* or *line power spectrum*.

### Functions with Even and Odd Symmetry

A function is said to be an *even function* if it possesses mirror symmetry about the origin, i.e., if $f(-t) = f(t)$. A function is said to be an *odd function* if it possesses negative symmetry about the origin, i.e., if $f(-t) = -f(t)$. It is obvious that the sum of two or more even functions will be an even function, and the sum of two or more odd functions will be an odd function. On the other hand, the product of two even or two odd functions is an even function, while the product of an even with an odd function is an odd function. If $f_e(t)$ is an even function,

$$\int_{-t_1}^{+t_1} f_e(t)\, dt = 2 \int_0^{+t_1} f_e(t)\, dt. \qquad (6.29)$$

and if $f_0(t)$ is an odd function,

$$\int_{-t_1}^{+t_1} f_0(t)\, dt = 0. \qquad (6.30)$$

By using these properties and the knowledge that cosine functions are even and sine functions are odd, we may easily show (see Problem 6.10) that an even function contains only cosine terms in its Fourier series (plus possibly a constant), and that an odd function contains only sine terms in its Fourier series. Hence, for an even function, the $\alpha_n$ are all real numbers, while for an odd function the $\alpha_n$ are all pure imaginaries. For any function that is neither even nor odd, at least some of the $\alpha_n$ must be complex.

### 6.4 APPROXIMATION OF PERIODIC FUNCTIONS BY A TRUNCATED FOURIER SERIES

The equality sign in equation 6.4 or equation 6.21 means that as more and more terms are added onto the summation of a finite number of terms of the series, the summation approaches $f(t)$ in the limit; e.g.,

$$f(t) = \lim_{k \to \infty} \sum_{n=-k}^{+k} \alpha_n \epsilon^{jn\omega_0 t}, \qquad (6.31)$$

where of course $n$ and $k$ are integers. If we truncate the series after $k$ terms, the resulting function, say $f_1(t)$, will not equal $f(t)$ but will be some approximation to $f(t)$. It is not clear, however, that the $\alpha_n$ coefficients, calculated in the usual way, will be the best set of coefficients to use in the truncated series, if we want $f_1(t)$ to approximate $f(t)$ in some sort of optimum way. Let us examine this question in more detail.

There are many ways of defining an error between $f_1(t)$ and $f(t)$. If we simply take the difference, $f_1(t) - f(t)$, we have an error that is a function of time. It is generally more useful to define an error that is a constant rather than time dependent. We do this by time averaging the error. If we average $[f_1(t) - f(t)]$, however, we have a situation where positive and negative errors cancel—thus this quantity may give a very deceptive measure of how well $f_1(t)$ fits $f(t)$. To avoid this problem, one might average the absolute value of the difference, or the difference squared. Of these two choices, the average-squared difference (mean-square error) is usually the easiest to handle mathematically and is the error criterion most often used in practice.

We shall now consider a truncated series

$$f_1(t) = \sum_{n=-k}^{+k} \beta_n \epsilon^{jn\omega_0 t}, \tag{6.32}$$

where the $\beta_n$ will be calculated in such a way as to minimize the mean-square error:

$$e \triangleq \frac{1}{T} \int_{-T/2}^{+T/2} [f(t) - f_1(t)]^2 \, dt$$

$$= \frac{1}{T} \int_{-T/2}^{+T/2} \left[ f(t) - \sum_{n=-k}^{+k} \beta_n \epsilon^{jn\omega_0 t} \right]^2 \, dt. \tag{6.33}$$

A necessary condition for this error to be minimized is

$$\frac{\partial e}{\partial \beta_m} = 0 \qquad \text{(for all } m\text{).} \tag{6.34}$$

(Obviously, no finite maximum of $e$ exists.) Differentiating equation 6.33 gives

$$\frac{\partial e}{\partial \beta_m} = \frac{1}{T} \int_{-T/2}^{+T/2} 2 \left[ f(t) - \sum_{n=-k}^{+k} \beta_n \epsilon^{jn\omega_0 t} \right] \epsilon^{jm\omega_0 t} \, dt = 0, \tag{6.35}$$

or

$$\int_{-T/2}^{+T/2} f(t) \epsilon^{jm\omega_0 t} \, dt = \int_{-T/2}^{+T/2} \sum_{n=-k}^{+k} \beta_n \epsilon^{j(n+m)\omega_0 t} \, dt. \tag{6.36}$$

If we integrate the right-hand side of this equation term by term, we obtain

$$\int_{-T/2}^{+T/2} f(t) \epsilon^{jm\omega_0 t} \, dt = \sum_{n=-k}^{+k} \beta_n \int_{-T/2}^{+T/2} \epsilon^{j(n+m)\omega_0 t} \, dt. \tag{6.37}$$

Since $\epsilon^{j(n+m)\omega_0 t} = \cos(n+m)\omega_0 t + j \sin(n+m)\omega_0 t$, and since the integral of both the cosine and sine functions over any integer number of periods is zero, we only get a contribution to the right-hand side of equation 6.37 for $n = -m$, so

$$\int_{-T/2}^{+T/2} f(t) \epsilon^{jm\omega_0 t} \, dt = T\beta_{(-m)}. \tag{6.38}$$

If we divide by $T$, the left-hand side of this equation is just $\alpha_{(-m)}$, from whence it follows that

$$\beta_m = \alpha_m. \tag{6.39}$$

In summary, if we calculate the Fourier coefficients as usual and form a truncated series, we obtain a sinusoidal series approximation to $f(t)$ which is optimum in a mean-square sense. It should also be noted that, as we seek to improve our approximation by adding more and more terms to the series, the previously calculated coefficients do not change and hence do not require recalculation.

### Convergence of Fourier Series at Discontinuities

In stating the Fourier theorem, we should have been slightly more precise, and have stated that for any periodic function satisfying the Dirichlet conditions, equations 6.4 and 6.21 are valid at all points at which $f(t)$ is continuous. The Dirichlet conditions allow functions with any finite number of jump discontinuities per period, and often (as in the case of our square wave) the function is not defined right at the discontinuity itself. The Fourier series, however, since it consists of continuous functions, will converge to some value at a point of discontinuity. If we examine the series for our square wave (equation 6.13), we note that at each jump point (i.e., $t = 1, 2, \ldots$) the series yields the value $f(t) = 0$, which is the midpoint between $f(t_i-)$ and $f(t_i+)$, $t_i$ being the times at which jumps occur. In fact, it may be shown that at any discontinuity in a function occurring at $t = t_i$, the series will always converge to the midpoint of the discontinuity, or

$$f(t_i) = \tfrac{1}{2}[f(t_i-) + f(t_i+)]. \tag{6.40}$$

### Gibbs' Phenomenon

The fact that a truncated Fourier series provides the best trigonometric approximation (for a given number of terms) to a function in a least-mean-square-error sense does not imply that the approximation point by point will be everywhere close. In fact, as one would expect, the slowest pointwise convergence of the series occurs near points of discontinuity in $f(t)$. This is illustrated in Figure 6.4, which shows a square wave approximated by two, three, and seven terms of the Fourier series. If we evaluate the maximum overshoot for each of these three approximations, we would find it to be

$$\delta f = 0.0895\Delta f, \tag{6.41}$$

where $\Delta f$ is the total discontinuity in $f(t)$ as indicated in the figure. As we add more and more terms to the truncated series, the point at which the maximum overshoot occurs moves closer and closer to the point of discontinuity, and the oscillation dies out sooner, but $\delta f$ remains unchanged! In fact, an over-

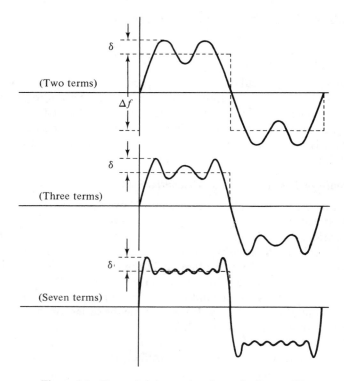

**Figure 6.4** Truncated Approximations of a Square Wave

shoot that is 8.95% of the jump occurs for any discontinuity in any waveform; this poor behavior of the truncated series near discontinuities is termed *Gibbs' phenomenon*.

### Uniqueness of the Fourier Series

Can two different functions, $f_1(t)$ and $f_2(t)$, have the same Fourier series? The answer is yes, provided that in each period the two functions differ from each other only at a discrete set of points. This is because the values of a function at any isolated set of points (mathematically, a set of measure zero) do not contribute to any integral of the function over an interval containing these points—hence our two functions will have identical Fourier coefficients. However, if we limit our discussion to functions that can be used to model physically generated signals, that is, signals of practical engineering usefulness, then the Fourier series is unique.

### Fourier Expansion of Nonperiodic Functions Over an Interval

If we have a nonperiodic function, we may represent it over any interval by a Fourier series, by treating that interval as a period and evaluating the

Fourier coefficients. Of course, outside the interval of interest the Fourier series is not correct; in fact, the series will give a periodically repeated version of the function as defined within the original interval. Within the original interval, however, the Fourier series representation of the function is perfectly valid.

## 6.5   CIRCUITS WITH PERIODIC FORCING FUNCTIONS

In Chapter 5 methods of analyzing circuits with sinusoidal forcing functions were discussed. The Fourier series allows us now to extend these methods to the analysis of circuits with periodic forcing functions, through the use of superposition.

If we consider a circuit with a single sinusoidal source, and use a phasor representation of that source, we know that the phasor representing any other quantity (output) in the circuit is obtained by multiplying the input phasor by an appropriate system function, $H(j\omega)$. That is, if the input is

$$x(t) = Xe^{j\omega t}$$

and the output is

$$y(t) = Ye^{j\omega t},$$

then

$$Y = H(j\omega)X.$$

If, however, $x(t)$ is not sinusoidal but is periodic and satisfies the Dirichlet conditions, we may represent it by its Fourier series (equation 6.21),

$$x(t) = \sum_{n=-\infty}^{+\infty} X_n e^{jn\omega_0 t}. \tag{6.42}$$

In equation 6.42 we have written the Fourier coefficient as $X_n$ (rather than $\alpha_n$) to emphasize its correspondence to the phasor $X$. In other words, the Fourier series simply represents a sum of sinusoidal inputs, and since the output phasor for each input component $X_n$ is just

$$Y_n = H(jn\omega_0)X_n, \tag{6.43}$$

the total output signal is just

$$y(t) = \sum_{n=-\infty}^{+\infty} H(jn\omega_0)X_n e^{jn\omega_0 t}. \tag{6.44}$$

Thus we may obtain the Fourier coefficients of any quantity in the circuit very easily, by simply multiplying the coefficients of the forcing function by the appropriate $H(jn\omega_0)$.

*Example 6.6*

In the circuit of Figure 6.5, the source voltage $v_1(t)$ is the square wave of Figure 6.1. The Fourier coefficients of the square wave were calculated (Example 6.5) as

$$V_{1_n} = \begin{cases} -\dfrac{j2}{n\pi} & (n = \pm 1, \pm 3, \pm 5, \dots) \\ 0 & (n = 0, \pm 2, \pm 4, \dots). \end{cases} \qquad (6.45)$$

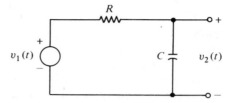

**Figure 6.5** Circuit of Example 6.6

The system function is

$$\frac{V_2}{V_1} = \frac{1}{1 + j\omega RC} = H(j\omega). \qquad (6.46)$$

Therefore, the Fourier coefficients of $v_2(t)$ are

$$V_{2_n} = \begin{cases} -\dfrac{j2}{n\pi} \dfrac{1}{1 + jn\pi RC} & (n = \pm 1, \pm 3, \pm 5, \dots) \\ 0 & (n = 0, \pm 2, \pm 4, \dots). \end{cases} \qquad (6.47)$$

(*Note:* In this example, $\omega_0 = \pi$.)

A comparison of the (double-sided) amplitude and power spectra of $v_1(t)$ and $v_2(t)$ is shown in Table 6.1. Note that $v_2(t)$ will be a smoother waveform than $v_1(t)$, since the high-frequency components of the Fourier series of $v_2(t)$ drop off in amplitude as $1/n^2$, while the high-frequency components of $v_1(t)$ drop off only as $1/n$. From knowledge of simple transient responses, can you sketch the waveform of $v_2(t)$?

**Table 6.1**[a]

|  | $v_1(t)$ | $v_2(t)$ |
|---|---|---|
| Amplitude spectrum | $\dfrac{2}{n\pi}$ | $\dfrac{2}{n\pi} \dfrac{1}{\sqrt{1 + (n\pi RC)^2}}$ |
| Power spectrum | $\dfrac{4}{n^2\pi^2}$ | $\dfrac{4}{n^2\pi^2} \dfrac{1}{1 + (n\pi RC)^2}$ |

[a]$n$ is odd.

Since the Fourier coefficients of any input and output quantity in a circuit are related by equation 6.43, if we denote amplitude spectra by $A$ and power spectra by $P$, it should be noted that

$$A_y = |H(jn\omega_0)|A_x \tag{6.48}$$

and

$$P_y = |H(jn\omega_0)|^2 P_x. \tag{6.49}$$

## 6.6  FOURIER TRANSFORMS

The Fourier series always converges to a periodic function, and, in fact, any partial sum of the series is a periodic function. It is possible, however, to extend the Fourier analysis so that it applies to nonperiodic functions; this is done by first considering a repeated (periodic) version of the nonperiodic function and then considering the limit as the period approaches infinity. Since we start with a Fourier series and then let the period approach infinity, we again must restrict ourselves to functions that satisfy the Dirichlet conditions, over the infinite interval. That is, our functions must be of bounded variation over the entire interval $-\infty < t < \infty$, and

$$\int_{-\infty}^{+\infty} |f(t)|\, dt < \infty. \tag{6.50}$$

For example, if we want a Fourier representation for a single pulse centered at the origin as in Figure 6.6, we may first consider the Fourier series of the entire pulse train as indicated by the dashed lines, and then let $T \to \infty$, so that in the limit only the single pulse at the origin remains in any finite interval.

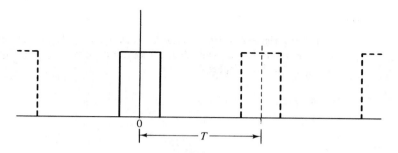

**Figure 6.6**  Repeated Version of a Pulse

### *Fourier Integral*

Let us define a discrete function $F(jn\omega_0)$ by multiplying both sides of equation 6.22 by $T$; i.e.,

$$T\alpha_n = \int_{-T/2}^{+T/2} f(t)\epsilon^{-jn\omega_0 t}\,dt \triangleq F(jn\omega_0). \tag{6.51}$$

Then we may rewrite equation 6.21 as

$$f(t) = \frac{1}{T} \sum_{n=-\infty}^{+\infty} F(jn\omega_0)\epsilon^{jn\omega_0 t}. \tag{6.52}$$

Note that since the harmonics, or spectral lines, occur at the frequencies $n\omega_0$, the separation or distance between adjacent lines is $\Delta\omega = \omega_0$, and $T = 2\pi/\omega_0 = 2\pi/\Delta\omega$. Thus, if we like, we may rewrite equation 6.52 as

$$f(t) = \frac{1}{2\pi} \sum_{n=-\infty}^{+\infty} F(jn\omega_0)\epsilon^{jn\omega_0 t}\,\Delta\omega. \tag{6.53}$$

Now, if we let $T$ increase, $\Delta\omega = 2\pi/T$ decreases; in other words, the spectral lines become closer or more dense, and in the limit we may consider the discrete frequencies $n\omega_0$ to "merge" into the continuous frequency variable $\omega$. Thus, in the limit, equation 6.53 is just the usual definition of an integral, and we may write

$$f(t) = \frac{1}{2\pi} \int_{-\infty}^{+\infty} F(j\omega)\epsilon^{j\omega t}\,d\omega$$

and, from equation 6.51,

$$F(j\omega) = \int_{-\infty}^{+\infty} f(t)\epsilon^{-j\omega t}\,dt. \tag{6.54}$$

These relationships, equations 6.54, are called Fourier integrals, and $f(t)$ and $F(j\omega)$ are said to be a *Fourier transform pair*.

### Example 6.7

Let us find the Fourier transform of a single rectangular pulse, of height $H$ and width $w$, centered at the origin (e.g., the pulse at the origin in Figure 6.2). Applying the Fourier integral, we obtain

$$\begin{aligned}
F(j\omega) &= \int_{-\infty}^{+\infty} f(t)\epsilon^{-j\omega t}\,dt = H\int_{-w/2}^{+w/2} \epsilon^{-j\omega t}\,dt \\
&= H\frac{1}{-j\omega}[\epsilon^{-j\omega t}]_{-w/2}^{+w/2} = H\frac{\epsilon^{j(\omega w/2)} - \epsilon^{-j(\omega w/2)}}{j\omega} \\
&= \frac{2H}{\omega}\sin\frac{\omega w}{2} = Hw\frac{\sin(\omega w/2)}{\omega w/2}. \tag{6.55}
\end{aligned}$$

Note that $F(j\omega)$ for this example is a purely real function of $\omega$—we would expect this from our Fourier series discussion and the fact that we are analyzing an even time function. As with the Fourier series, an odd time function will yield a purely imaginary $F(j\omega)$, while an $f(t)$ that is neither odd nor even will yield a complex $F(j\omega)$.

If we sketch $F(j\omega)$ for this example, we get the continuous function indicated in Figure 6.7.

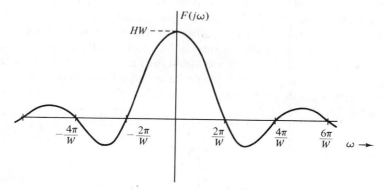

**Figure 6.7** $F(j\omega)$ of Rectangular Pulse Centered at Origin

### Spectral Interpretation of $F(j\omega)$

The Fourier transform $F(j\omega)$ is a continuous (complex) function of $\omega$, and we may plot $|F(j\omega)|$, $\underline{/F(j\omega)}$, $|F(j\omega)|^2$, etc., just as we did for the Fourier series coefficients, $\alpha_n$. Let us examine the interpretation of these plots, which we may call *continuous spectra*, in contrast to the line or discrete spectra resulting from a Fourier series analysis.

Both the Fourier series and Fourier integral involve representing a time function by an infinite sum of sinusoids; in the Fourier integral, however, the frequencies involved fill up the entire frequency line; i.e., all frequencies are included. If we compare the Fourier series

$$f(t) = \sum_{n=-\infty}^{+\infty} \alpha_n \epsilon^{jn\omega_0 t}$$

and Fourier integral

$$f(t) = \frac{1}{2\pi} \int_{-\infty}^{+\infty} F(j\omega)\epsilon^{j\omega t}\, d\omega,$$

we see that the quantity in the Fourier integral which corresponds to $\alpha_n$ is

$$\frac{1}{2\pi}F(j\omega)\, d\omega, \tag{6.56}$$

and this is a quantity having a differential magnitude. This means that the amplitude of each sinusoidal component in the Fourier integral is infinitesimal. None the less, a plot of $F(j\omega)$ like that in Figure 6.7 still shows the relative importance of the contributions of various frequency intervals (bands) to the total $f(t)$. If, for example, we pass our rectangular pulse through a circuit that will pass only frequencies up to $2\pi/w$ rad/s, we should expect a highly distorted version of our pulse at the output. On the other hand, if the circuit passes a wide band of frequencies, such that nearly all the area under the $F(j\omega)$ curve is in the pass band, then a much more faithful reproduction of the pulse will appear at the output. [In Example 6.7, $F(j\omega)$ is real; in general, to completely specify $F(j\omega)$, we would need a plot of both its magnitude and angle, and a plot of $|F(j\omega)|$ would be called an *amplitude spectrum*, by analogy to the Fourier series. For Example 6.7, a plot of $|F(j\omega)|$ appears as in Figure 6.8.]

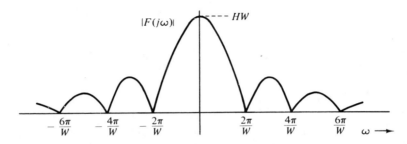

**Figure 6.8**   Amplitude Spectrum of Rectangular Pulse

In previous discussions we termed $f^2(t)$ the *instantaneous power* associated with a signal $f(t)$. For periodic signals, we also found the time-average power by averaging $f^2(t)$ over a period; i.e.,

$$P = \frac{1}{T} \int_T f^2(t)\, dt. \tag{6.57}$$

Since power is (dimensionally) energy divided by time, the integral of $f^2(t)$ over any interval $T$ is termed the energy in $f(t)$ over that interval, and

$$E \triangleq \int_{-\infty}^{+\infty} f^2(t)\, dt \tag{6.58}$$

is called the *total energy* of the signal $f(t)$. For a periodic signal, the total energy is of course infinite. For a nonperiodic signal amenable to Fourier integral analysis, however, the Dirichlet conditions imply that

$$E = \int_{-\infty}^{+\infty} f^2(t)\, dt < \infty. \tag{6.59}$$

(For the pulse of Example 6.7, the total energy is $wH^2$.) For such a finite energy signal, the time-average power is zero, since the interval is infinite but the total energy is finite. Thus periodic signals are members of a class of signals that are often called *power signals*; they have finite (nonzero) time-average power but infinite total energy. In contrast, signals that have Fourier integrals belong to a class called *energy signals*; they have zero time-average power but finite total energy. Therefore, in Fourier series analysis we are dealing with power signals, and in Fourier integral analysis we are dealing with energy signals.

If we use a Fourier integral representation for $f(t)$ in equation 6.58 and interchange the order of integration, we obtain

$$
\begin{aligned}
E &= \int_{-\infty}^{+\infty} f(t)f(t)\, dt = \int_{-\infty}^{+\infty} f(t)\left[\frac{1}{2\pi}\int_{-\infty}^{+\infty} f(j\omega)\epsilon^{j\omega t}\, d\omega\right] dt \\
&= \frac{1}{2\pi}\int_{-\infty}^{+\infty} F(j\omega)\left[\int_{-\infty}^{+\infty} f(t)\epsilon^{j\omega t}\, dt\right] d\omega \\
&= \frac{1}{2\pi}\int_{-\infty}^{+\infty} F(j\omega)F(-j\omega)\, d\omega = \frac{1}{2\pi}\int_{-\infty}^{+\infty} |F(j\omega)|^2\, d\omega \\
&= \int_{-\infty}^{+\infty} |F(jf)|^2\, df. \qquad\qquad [\omega = 2\pi f]
\end{aligned}
\tag{6.60}
$$

Equation 6.60, which states that the total energy can be obtained either by integrating $f^2(t)$ over all time or $|F(jf)|^2$ over all frequencies, is one form of *Parseval's theorem*. Thus it appears that if we want to know how much energy is contributed to $f(t)$ by a certain frequency band, we should integrate $(1/2\pi)|F(j\omega)|^2$ over that band of $\omega$, or $|F(jf)|^2$ over the corresponding band of $f$. For this reason, a plot of $(1/2\pi)|F(j\omega)|^2$ versus $\omega$, or a plot of $|F(jf)|^2$ versus $f$, is called an *energy density spectrum*. The integral of the energy density spectrum over all frequencies is simply the total energy of the signal. The energy density spectrum for the pulse of Example 6.7 is sketched (as a function of $f$) in Figure 6.9.

### Some Properties of Fourier Transforms

With the exception of functions that differ from each other only at isolated points, the Fourier integral is unique, and every $f(t)$ that satisfies the Dirichlet conditions corresponds to a Fourier integral or transform; i.e.,

$$ F(j\omega) = \mathfrak{F}[f(t)], $$

where $\mathfrak{F}$ denotes the linear operator which weights $f(t)$ with $\epsilon^{-j\omega t}$ and integrates the product over all time. $f(t)$ and $F(j\omega)$ thus form a unique pair of functions, called a *Fourier transform pair*. Let us now examine some important properties of Fourier transform pairs.

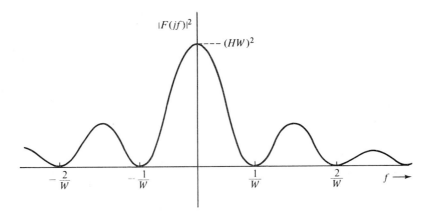

**Figure 6.9**  Energy Spectral Density of Rectangular Pulse

1. If $f(t)$ is an even function, $F(j\omega)$ is real.
2. If $f(t)$ is an odd function, $F(j\omega)$ is imaginary.

Since

$$F(j\omega) = \int_{-\infty}^{+\infty} f(t)\epsilon^{-j\omega t} \, dt$$

$$= \int_{-\infty}^{+\infty} f(t) \cos \omega t \, dt - j \int_{-\infty}^{+\infty} f(t) \sin \omega t \, dt$$

$$= \operatorname{Re} F(j\omega) + j \operatorname{Im} F(j\omega), \tag{6.61}$$

properties 1 and 2 follow from the properties of even and odd functions. (How?)

3. The real part of $F(j\omega)$ is an even function of $\omega$.
4. The imaginary part of $F(j\omega)$ is an odd function of $\omega$.

(From equation 6.61, how do these properties follow?)

5. $|F(j\omega)|$ is an even function of $\omega$.
6. $\underline{/F(j\omega)}$ is an odd function of $\omega$.

(See Problem 6.18.)

7. $\mathfrak{F}[af(t)] = aF(j\omega)$, where $a$ is any constant. (Amplitude scaling.) (*Note*: $a$ simply comes outside the integral sign.)
8. $\mathfrak{F}[f(at)] = (1/|a|)F(j\omega/a)$. (Time scaling.)

We may write

$$\mathcal{F}[f(at)] = \int_{-\infty}^{+\infty} f(at)\epsilon^{-j\omega t} \, dt. \tag{6.62}$$

We now change variables by letting $\tau = at$, or $dt = (1/a) \, d\tau$. If $a$ is positive, we have

$$\mathcal{F}[f(at)] = \frac{1}{a} \int_{-\infty}^{+\infty} f(\tau)\epsilon^{-(j\omega/a)\tau} \, d\tau = \frac{1}{a} F\left(\frac{j\omega}{a}\right). \tag{6.63}$$

If $a$ is negative, we have

$$\mathcal{F}[f(at)] = \frac{1}{a} \int_{+\infty}^{-\infty} f(\tau)\epsilon^{-(j\omega\tau/a)} \, d\tau$$

$$= -\frac{1}{a} \int_{-\infty}^{+\infty} f(\tau)\epsilon^{-(j\omega\tau/a)} \, d\tau = -\frac{1}{a} F\left(\frac{j\omega}{a}\right). \tag{6.64}$$

Hence property 8 follows.

9. $\mathcal{F}[f(t - \tau)] = F(j\omega)\epsilon^{-j\omega\tau}.$ (Time-shift property)

*Proof:*

$$\mathcal{F}[f(t - \tau)] = \int_{-\infty}^{+\infty} f(t - \tau)\epsilon^{-j\omega t} \, dt. \tag{6.65}$$

Let $t - \tau = \zeta$, $dt = d\zeta$. Then

$$\mathcal{F}[f(t - \tau)] = \int_{-\infty}^{+\infty} f(\zeta)\epsilon^{-j\omega(\zeta+\tau)} \, d\zeta$$

$$= \epsilon^{-j\omega\tau} \int_{-\infty}^{+\infty} f(\zeta)\epsilon^{-j\omega\zeta} \, d\zeta = F(j\omega)\epsilon^{-j\omega\tau}. \tag{6.66}$$

10. $\mathcal{F}[af_1(t) + bf_2(t)] = aF_1(j\omega) + bF_2(j\omega).$ (Superposition property)

(This follows from the fact that the Fourier integral is a linear operation. Can you show this?)

11. $\mathcal{F}\left[\dfrac{df(t)}{dt}\right] = (j\omega)F(j\omega).$ (Differentiation property)

12. $\mathcal{F}\left[\displaystyle\int_{-\infty}^{t} f(t) \, dt\right] = \dfrac{F(j\omega)}{j\omega}.$ (Integration property)

(See Problem 6.19.)

13. $\mathcal{F}[f(t)\epsilon^{j\omega_0 t}] = F[j(\omega - \omega_0)].$  (Modulation property)

(See Problem 6.20.)

14. $\mathcal{F}[f(t) \cos \omega_0 t] = \frac{1}{2}F[j(\omega - \omega_0)] + \frac{1}{2}F[j(\omega + \omega_0)].$
(Real modulation)

(This follows from property 13 if we express $\cos \omega_0 t$ in exponential form.)

### 6.7 SPECTRAL RELATIONSHIPS IN NETWORKS

*Input–Output Fourier Transform Relationship*

In Section 6.5 we saw that if the input to a network is periodic, with Fourier coefficients $X_n$, then the Fourier coefficients $Y_n$ of the output are given by (equation 6.43)

$$Y_n = H(jn\omega_0)X_n.$$

Since the Fourier transform was derived as a limit of the Fourier series as $T \longrightarrow \infty$, one would suspect that a similar relationship holds for the input and output Fourier transforms. We may show that this is indeed the case by going back to the differential equation relating the output $y(t)$ and the input $x(t)$. We may write this equation in operator form (as we did in Chapter 5):

$$M(D)y(t) = N(D)x(t), \tag{6.67}$$

where

$$M(D) = a_n \frac{d^n}{dt^n} + a_{n-1} \frac{d^{n-1}}{dt^{n-1}} + \ldots + a_0 \tag{6.68}$$

$$N(D) = b_m \frac{d^m}{dt^m} + b_{m-1} \frac{d^{m-1}}{dt^{m-1}} + \ldots + b_0. \tag{6.69}$$

From the properties of Fourier transforms, we know that (property 11)

$$\mathcal{F}\left[\frac{d^n f(t)}{dt^n}\right] = (j\omega)^n F(j\omega). \tag{6.70}$$

Hence, if we take the Fourier transform of equation 6.67 and make use of properties 10 and 11, we have

$$M(j\omega)Y(j\omega) = N(j\omega)X(j\omega), \tag{6.71}$$

where $Y(j\omega)$ and $X(j\omega)$ are the Fourier transforms of $y(t)$ and $x(t)$, respectively. Thus

$$Y(j\omega) = \frac{N(j\omega)}{M(j\omega)} X(j\omega) = H(j\omega)X(j\omega), \qquad (6.72)$$

as we would expect from equation 6.43. This then gives another interpretation of the system function $H(j\omega)$; it is not only the ratio of the output and input phasors at any specific frequency $\omega$, but it is also the ratio of the output and input Fourier transforms, for any functions for which these transforms exist.

A word of caution is in order here. When we derived equations 6.43 and 6.72 we (tacitly) ignored any initial conditions that would arise if specific stored energies in the network were specified at any given time epoch. This is because $H(j\omega)$ is essentially derived from and related to the sinusoidal steady state, in which initial condition specifications are not appropriate. This was perfectly clear until we arrived at equation 6.72, which can be used, for example, to find the transient response of a network to a pulse-type input. This would be done by first finding the transform of the input, then multiplying by $H(j\omega)$, and finally taking the inverse transform to get $y(t)$. By this method, however, we obtain only the response of the network to the input signal, not the part of the response due to any specified initial conditions. (The initial condition response can of course be evaluated separately by solving the corresponding homogeneous equation.) It should be kept in mind that the Fourier transform, even though it is used to represent transient waveforms, none the less represents these waveforms as a sum of steady-state sinusoids (albeit of infinitesimal amplitude). The finding of both parts of a network response, i.e., the initial condition response and the response due to the input function, is discussed at length in Chapters 7 and 8.

### Example 6.8

In the circuit of Figure 6.10 we will let

$$v_1(t) = \begin{cases} \epsilon^{-\alpha t} & (t \geq 0) \\ 0 & (t < 0), \end{cases} \qquad (6.73)$$

and we will find $v_2(t)$ by using Fourier transforms. [Note that the Fourier transform of $v_1(t)$ exists, but that the transform of $\epsilon^{-\alpha t}$ for all $t$ does not—why?]

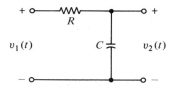

**Figure 6.10**   Circuit of Example 6.8

If we denote the transform of $v_1(t)$ by $V_1(j\omega)$, we have

$$V_1(j\omega) = \int_{-\infty}^{+\infty} v_1(t)\epsilon^{-j\omega t} \, dt = \int_0^\infty \epsilon^{-(\alpha+j\omega)t} \, dt$$

$$= -\frac{1}{\alpha+j\omega}[\epsilon^{-(\alpha+j\omega)t}]_0^\infty = \frac{1}{\alpha+j\omega}. \tag{6.74}$$

[Note that $\epsilon^{-(\alpha+j\omega)t} = \epsilon^{-\alpha t}(\cos \omega t - j \sin \omega t)$ does indeed approach zero as $t \longrightarrow \infty$.] Thus we have derived a Fourier transform pair:

$$\left.\begin{matrix} \epsilon^{-\alpha t} & (t \geq 0) \\ 0 & (t < 0) \end{matrix}\right\} \longleftrightarrow \frac{1}{j\omega + \alpha}. \tag{6.75}$$

The system function, $H(j\omega)$, we have found previously; it is

$$H(j\omega) = \frac{1/RC}{j\omega + (1/RC)}. \tag{6.76}$$

The output voltage transform $V_2(j\omega)$ is therefore

$$V_2(j\omega) = H(j\omega)V_1(j\omega) = \frac{1/RC}{(j\omega + \alpha)[j\omega + (1/RC)]}. \tag{6.77}$$

We may find the inverse transform of this function directly, but it is easier if we first express $V_2(j\omega)$ as the sum of two terms,

$$V_2(j\omega) = \frac{C_1}{j\omega + \alpha} + \frac{C_2}{j\omega + (1/RC)}, \tag{6.78}$$

where $C_1$ and $C_2$ are constants. (This is called a *partial fraction expansion*, and is discussed in a general way in Chapter 8.) We may evaluate $C_1$ and $C_2$ by putting the terms of equation 6.78 over the common denominator and comparing the numerator with that of equation 6.77. This gives

$$C_1 = \frac{1}{1 - \alpha RC}, \qquad C_2 = \frac{-1}{1 - \alpha RC}.$$

Since the Fourier transform is a linear operation and we already know the general transform pair of equation 6.75, we may write immediately

$$v_2(t) = \begin{cases} C_1\epsilon^{-\alpha t} + C_2\epsilon^{-t/RC} & (t \geq 0) \\ \dfrac{1}{1 - \alpha RC}[\epsilon^{-\alpha t} - \epsilon^{-(t/RC)}] & (t \geq 0). \end{cases} \tag{6.79}$$

To work Example 6.8 we needed only the single transform pair of equation 6.75. Extensive tables of Fourier transform pairs have been published and may be used to find the particular integral solutions to linear differential equations, just as we did in Example 6.8. Using transforms to solve differential equations is analogous to using logarithms to solve arithmetic problems; one first transforms the original problem variables into other variables, in terms of which the solution operations are simpler—after solving the problem in the new domain, one then transforms back to the original one. In using Fourier transforms, we start with differential equations in the time domain, convert our functions to a frequency-domain representation, solve the resulting algebraic equations, and finally convert our solution back to the time domain. A similar process is used with Laplace transforms, which are discussed in Chapter 8.

### *Input–Output Energy Spectral Density Relationship*

Let us denote the energy spectral density of a signal as (from equation 6.60)

$$S(\omega) = \frac{1}{2\pi} |F(j\omega)|^2, \tag{6.80}$$

and the total energy in the signal is

$$E = \int_{-\infty}^{+\infty} S(\omega) \, d\omega. \tag{6.81}$$

Since the input and output transforms of a network are related by

$$Y(j\omega) = H(j\omega) X(j\omega), \tag{6.82}$$

it follows that

$$|Y(j\omega)| = |H(j\omega)| |X(j\omega)|, \tag{6.83}$$

and hence

$$S_y(\omega) = |H(j\omega)|^2 S_x(\omega). \tag{6.84}$$

Therefore, if we know the energy spectral density of the input signal, we may find the energy spectral density of the output directly, without calculating the output time function.

### *Example 6.9*

The energy spectral density of a rectangular pulse of height $H$ and width $w$ may be found by squaring the magnitude of the transform of equation 6.55:

$$S(\omega) = \frac{H^2 w^2}{2\pi} \frac{\sin^2(\omega w/2)}{(\omega w/2)^2} \tag{6.85}$$

(Since the energy spectral density is not dependent on $/\mathcal{F}(j\omega)$, the location of the pulse along the time axis is immaterial. This may also be seen from the time-shift property (property 9) of the Fourier transform; i.e.,

$$|\mathfrak{F}[f(t - \tau)]|^2 = |F(j\omega)|^2|\epsilon^{-j\omega\tau}|^2 = |F(j\omega)|^2.) \tag{6.86}$$

This function, $S(\omega)$, has the general form of the sketch of Figure 6.9.

Let us suppose that our rectangular pulse is $v_1(t)$ in the circuit of Figure 6.10. The energy spectral density $S_2(\omega)$ of the output voltage can then be written immediately:

$$S_2(\omega) = |H(j\omega)|^2 S_1(\omega)$$

$$= \frac{H^2 w^2}{2\pi(1 + R^2 C^2 \omega^2)} \frac{\sin^2(\omega w/2)}{(\omega w/2)^2}. \tag{6.87}$$

Note that the output energy spectral density falls off much more rapidly than does the input energy density with increasing frequency, again illustrating the low-pass nature of this circuit.

### Fourier Transforms and the Transmission of Signals

We have seen that in general the Fourier transform of a signal (and therefore the signal itself), and also the energy spectral density of a signal, get modified as the signal is passed through a network. In many cases, however, one would like to "get the signal through" a particular circuit, or transmit if from one place to another via a transmission circuit, with as little change in the signal as possible—at least with only a small alteration in the shape or waveform. Let us look at what the frequency characteristics of a network must be in order to accomplish this.

A circuit or transmission system is said to be *distortionless* if, for an input signal $x(t)$, the output signal is

$$y(t) = Ax(t - \tau). \tag{6.88}$$

In other words, the signal may be delayed in time by an amount $\tau$ seconds, and scaled in amplitude by a factor $A$, but the shape of the signal remains unchanged. If we make use of the amplitude scaling and time-shift properties of Fourier transforms, equation 6.88 transforms as

$$Y(j\omega) = AX(j\omega)\epsilon^{-j\omega\tau}, \tag{6.89}$$

which corresponds to a system function

$$H(j\omega) = A\epsilon^{-j\omega\tau}. \tag{6.90}$$

Since

$$|H(j\omega)| = A \tag{6.91}$$

and

$$\underline{/H(j\omega)} = -\omega\tau, \tag{6.92}$$

we see that the system function for a distortionless network must have a flat or constant amplitude characteristic and a linear phase characteristic, i.e., a phase characteristic that decreases linearly with increasing $\omega$.

Any network that is characterized by only resistance and transmission time delays will be a distortionless network. We know from the characteristics of physical networks, however, that any network, even if it is designed to essentially serve as a transmission line, will have inductance and capacitance associated with it; it is not possible to build a network having exactly the system function of equation 6.90. Let us look at one consequence of not being able to physically realize equation 6.90 exactly.

In Example 6.7 we found the transform of a rectangular pulse of height $H$ and width $w$ to be

$$F(j\omega) = Hw\,\frac{\sin(\omega w/2)}{\omega w/2}. \tag{6.93}$$

From this we can see that as the time pulse gets narrow, the spectrum spreads out, and vice versa. For example, if we define a *bandwidth* for this spectrum as the width of the main lobe (see Figure 6.7), we get a bandwidth

$$B = \frac{4\pi}{w}, \tag{6.94}$$

so that

$$(\text{pulse duration})(\text{bandwidth}) = (\text{constant}) = 4\pi. \tag{6.95}$$

Any bandwidth that we choose to define for the spectrum will yield the same result; i.e., the pulse duration multiplied by the bandwidth is constant. Let us examine this property with another kind of pulse, this time, one that is not time-limited.

### Example 6.10

Consider the Gaussian-shaped pulse

$$f(t) = \epsilon^{-a^2 t^2}, \tag{6.96}$$

which is sketched in Figure 6.11(a). The Fourier transform of this pulse is

$$F(j\omega) = \int_{-\infty}^{+\infty} \epsilon^{-a^2 t^2} \epsilon^{-j\omega t}\, dt = \int_{-\infty}^{+\infty} \epsilon^{-a^2 t^2} \cos \omega t\, dt$$

$$= \frac{\sqrt{\pi}}{a} \epsilon^{-\omega^2/4a^2}. \tag{6.97}$$

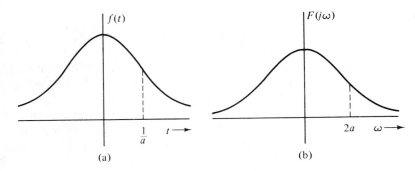

**Figure 6.11**  Gaussian-shaped Pulse and Its Transform

Note again that as the pulse is made more narrow (by increasing $a$), the spectrum becomes wider, and vice versa. For example, we might define the width $W$ of the pulse in this case to be the interval over which $f(t)$ is greater than $1/\epsilon$ times its value at $t = 0$; i.e., $W = 2T$, where

$$\epsilon^{-a^2T^2} = \epsilon^{-1}, \tag{6.98}$$

or

$$W = \frac{2}{a}. \tag{6.99}$$

Similarly, if we define the bandwidth $B$ as the interval over which $F(j\omega)$ is greater than $1/\epsilon$ times its value at $\omega = 0$, we have $B = 2\omega_1$, where

$$\epsilon^{-\omega_1^2/4a^2} = \epsilon^{-1},$$

or

$$B = 4a. \tag{6.100}$$

Thus the pulse duration–bandwidth product is

$$BW = 8. \tag{6.101}$$

Example 6.10 exemplifies a result that can be proved in general—that no matter how we define the pulse width and bandwidth, the two vary inversely with each other, and their product is constant. Hence we may state that the bandwidth of a network required to transmit a pulse is inversely proportional to the pulse width. This idea is further exemplified and amplified in more advanced texts dealing with Fourier analysis.[2]

It should be evident from the discussions in this chapter that the Fourier series and Fourier transform play a very prominent role in network theory—

[2]See, for example, A. Papoulis, *The Fourier Integral and Its Applications*, McGraw-Hill Book Company, New York, 1962.

particularly in the analysis and design of communication circuits. In Chapter 8 we shall discuss another type of transform, called the *Laplace transform,* which is closely related to the Fourier transform but is somewhat more convenient for the transient analysis of circuits, for reasons that we shall see.

## PROBLEMS

**6.1** Verify the orthogonality properties of sinusoids indicated in equations 6.6.

**6.2** Expand the function indicated in Figure 6.12 in a Fourier series, and show both the sine–cosine and exponential forms of the series.

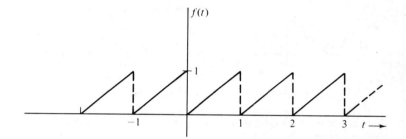

**Figure 6.12**   Waveform of Problem 2

**6.3** Repeat Problem 6.2 for the function shown in Figure 6.13.

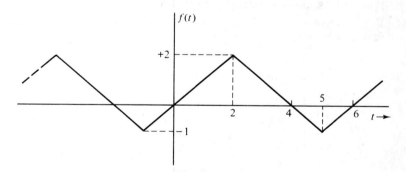

**Figure 6.13**   Waveform of Problem 3

**6.4** A sinusoid is full-wave-rectified to give the function shown in Figure 6.14. Repeat Problem 6.2 for this function.

**6.5** Repeat Problem 6.2 for the function shown in Figure 6.15.

**6.6** Derive equation 6.22 by using the integral expressions for $a_k$ and $b_k$ and Euler's relationships.

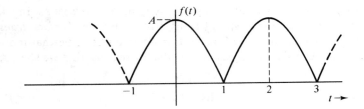

**Figure 6.14** Waveform of Problem 6.4

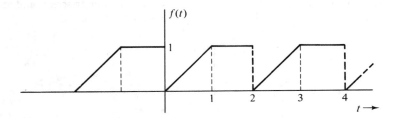

**Figure 6.15** Waveform of Problem 6.5

**6.7** Find expressions for, and sketch, the amplitude, phase, and power spectra for the waveform of Problem 6.2.

**6.8** Repeat Problem 6.7 for the waveform of Problem 6.4.

**6.9** Verify equation 6.27.

**6.10** Show that an even function has only cosine terms in its Fourier series expansion, and that an odd function has only sine terms.

**6.11** If the current source in Figure 6.16 generates the waveform of Problem 6.2, find the Fourier coefficients of the voltage $v(t)$.

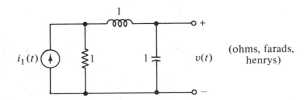

**Figure 6.16** Circuit of Problem 6.11

**6.12** Find an expression for the power spectrum of $v(t)$ in Problem 6.11.

**6.13** Find the Fourier transform of the function shown in Figure 6.17.

**6.14** Find the Fourier transform of the function

$$f(t) = \begin{cases} \sin t & 0 \leq t \leq 2\pi \\ 0 & \text{elsewhere} \end{cases}$$

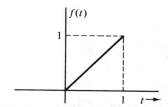

**Figure 6.17**  Waveform of Problem 6.13

**6.15**  Find the Fourier transform of the function

$$f(t) = \begin{cases} \epsilon^{-t} & t \geq 0 \\ 0 & t < 0. \end{cases}$$

**6.16**  For the functions of Problems 6.13 and 6.15, find expressions for (and sketch) the energy density spectra.

**6.17**  From equation 6.61 show that the real part of $F(j\omega)$ is an even function of $\omega$ and that the imaginary part of $F(j\omega)$ is an odd function of $\omega$.

**6.18**  Show that $|F(j\omega)|$ is an even function of $\omega$ and that $\underline{/F(j\omega)}$ is an odd function of $\omega$.

**6.19**  Verify the differentiation and integration properties (properties 11 and 12) of Fourier transforms.

**6.20**  Verify the modulation property (property 13) of Fourier transforms, and show that property 14 follows.

**6.21**  If the current source in Figure 6.16 generates a current pulse as indicated in Figure 6.17, find the energy spectral density of $v(t)$.

# 7

# Impulse Response, Convolution, and Digital Computer Simulation of Linear Networks

In this chapter we shall be concerned with obtaining time-domain solutions for networks with general forcing functions, and putting these solutions into a form amenable to digital analysis. First, however, we must consider a very important but rather peculiar time function called the *Dirac delta function* or simply *delta function*. *Impulse function* is another term that means the same thing, and we shall use impulse function and delta function interchangably.

## 7.1 THE IMPULSE FUNCTION AND ITS PROPERTIES

Let us consider a rectangular pulse of height $H$ and width $W$, as indicated in Figure 7.1(a). The area of this pulse is $HW$. If we double $H$ and decrease $W$ by a factor of 2, the area remains unchanged. In fact, we may repeat this process as many times as we wish, and the area will be unchanged. If we start with a pulse of unit area, then in the limit the height of the pulse is unbounded, its width becomes infinitesimal, but the area always remains unity. This limiting function is the *unit impulse* or *delta function* and is described by the following characteristics:

$$\delta(t - \tau) = 0 \qquad (t \neq \tau) \qquad (7.1)$$

$$\delta(t - \tau) \longrightarrow \infty \qquad (t = \tau) \qquad (7.2)$$

$$\int_{-\infty}^{+\infty} \delta(t - \tau)dt = 1. \qquad (7.3)$$

(In the ordinary mathematical sense, the three characteristics just listed are

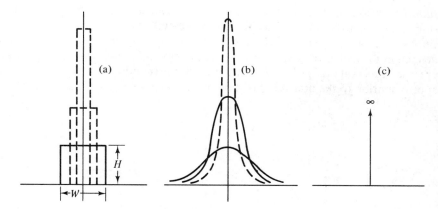

**Figure 7.1**  Successive Approximations to an Impulse

not consistent, since the integral of any function that is nonzero at only a single isolated point is zero, no matter how large the value of the function at that point becomes. Hence these three properties of the impulse function should not be taken as a rigorous mathematical definition but rather as an indication of a limit that is approached by the finite pulse. The use of impulse functions can be rigorously justified by the use of the *theory of distributions*,[1] but here we will encounter no difficulties if we always treat the impulse function as a limit approached by a perfectly well-defined finite pulse.)

It is not necessary to consider the impulse function to be a limiting case of a rectangular pulse—we may start with any shape pulse that we choose. For example, we may let

$$f(t) = \frac{a}{\sqrt{\pi}} \epsilon^{-a^2 t^2} \tag{7.4}$$

and consider the limit as $a \rightarrow \infty$. [The integral of this function over the interval $(-\infty, +\infty)$ is unity.] This approximation for the impulse function is indicated in Figure 7.1(b), and there are many other pulse shapes that we could choose. The usual symbol for an impulse is shown in Figure 7.1(c).

There is another interpretation that we may give to the unit impulse function. Its indefinite integral is the unit step function. That is, from the properties indicated by equations 7.1 through 7.3, it is clear that

$$\int_{-\infty}^{t} \delta(t - \tau) \, dt = u(t - \tau). \tag{7.5}$$

Alternatively, we may think of the impulse as the derivative of the unit step

---

[1]See, for example, B. Friedman, *Principles and Techniques of Applied Mathematics*, John Wiley & Sons, Inc., New York, 1956.

function, which has zero slope everywhere except at one point, where the slope is infinite. Again, of course, this statement is mathematically poor, since at a point of discontinuity the derivative, strictly speaking, cannot be defined in the usual way. Again, however, we may think in terms of a limiting process; a rectangular pulse approximation to the impulse corresponds to the approximation to the unit step function as indicated in Figure 7.2.

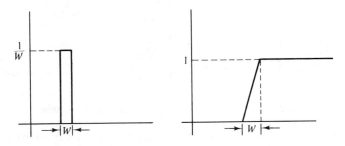

**Figure 7.2** Integral of Rectangular Pulse

In most practical applications, as we shall see, the impulse function is multiplied by some other function and the product is integrated. Suppose that a rectangular pulse is multiplied by a continuous function and the product integrated. As the pulse is made to approach an impulse, it becomes (in the limit) sufficiently narrow that over its duration the continuous function $f(t)$ may be considered constant, as indicated in Figure 7.3. Thus, in the limit, $f(\tau)$ may be taken outside of the integral, and we have

$$\int_{-\infty}^{+\infty} f(t)\,\delta(t-\tau)\,dt = f(\tau) \int_{-\infty}^{+\infty} \delta(t-\tau)\,dt = f(\tau). \qquad (7.6)$$

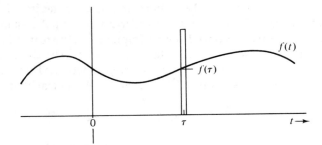

**Figure 7.3** Product of Continuous Function and Rectangular Pulse

Thus the value of the integral is just the value of $f(t)$ at the epoch at which the impulse function occurs, and, as a special case,

$$\int_{-\infty}^{+\infty} f(t)\,\delta(t)\,dt = f(0). \tag{7.7}$$

### Fourier Transform of the Impulse Function

Strictly speaking, we should say that the impulse function does not possess a Fourier transform, at least in the usual sense, since it does not satisfy the Dirichlet conditions. (Why?) However, we may again consider a limiting process, and define the Fourier transform of the impulse as the limit approached by the transform of a finite pulse, as that pulse approaches an impulse. For example, we have seen that the transform of a rectangular pulse of height $H$ and width $w$ centered at the origin is

$$F(j\omega) = Hw\,\frac{\sin\,(w\omega/2)}{w\omega/2}. \tag{7.8}$$

If we now let $H$ increase and $w$ decrease in such a way that $Hw = 1$, then $F(j\omega) \longrightarrow 1$ for any finite value of $\omega$ (as we may easily show by applying l'Hospital's rule). Formally, using the property of equation 7.6, we may write

$$\int_{-\infty}^{+\infty} \delta(t)\epsilon^{-j\omega t}\,dt = \epsilon^{-j\omega 0} = 1. \tag{7.9}$$

If the impulse is not at the origin, then

$$\int_{-\infty}^{+\infty} \delta(t - \tau)\epsilon^{-j\omega t}\,dt = \epsilon^{-j\omega \tau}. \tag{7.10}$$

Thus we see that the impulse function has an amplitude and energy spectral density which are constant—all frequencies, so to speak, are equally important, and all bands of frequencies of a fixed width contribute equally to the energy of the impulse, regardless of where those bands are located on the frequency axis.

### Impulse Response and System Function

If we have a circuit or system that is initially relaxed (no initial conditions), and if the input or forcing function is an impulse, then the corresponding output is called the *impulse response* of the system. If we denote the system input by $x(t)$ and the output by $y(t)$, then in the transform domain we have

$$Y(j\omega) = H(j\omega)X(j\omega) = H(j\omega)\cdot 1 = H(j\omega) \tag{7.11}$$

if $x(t) = \delta(t)$. Since $y(t)$ in this case is the impulse response, which we will denote by $h(t)$, we conclude (from equation 7.11) that the impulse response and the system function are a Fourier transform pair; i.e.,

$$H(j\omega) = \mathfrak{F}[h(t)]. \qquad (7.12)$$

### Impulse Response of Some Simple Circuits

#### Example 7.1

We will find the impulse response of the circuit of Figure 7.4, where the input is the terminal voltage and the output is the current. The appropriate system function is

$$H(j\omega) = \frac{I(j\omega)}{V(j\omega)} = \frac{1}{Z(j\omega)} = \frac{1}{R + j\omega L} = \frac{1}{L} \frac{1}{j\omega + (R/L)}. \qquad (7.13)$$

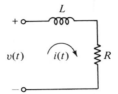

**Figure 7.4**  Circuit of Example 7.1

We have already encountered this transform in Chapter 6; it is the transform of an exponential for positive time; i.e.,

$$
h(t) = \text{(current impulse response)}
$$

$$
= \begin{cases} \dfrac{1}{L}\,\epsilon^{-(R/L)t} & (t \geq 0) \\ 0 & (t > 0). \end{cases} \qquad (7.14)
$$

[Alternatively, we may write

$$h(t) = \frac{1}{L}\,\epsilon^{-(R/L)t}u(t).] \qquad (7.15)$$

#### Example 7.2

We will find the impulse response of the circuit of Figure 7.5, where the input is an impulse of current and the output is the terminal voltage.

$$
H(j\omega) = \frac{V(j\omega)}{I(j\omega)} = \frac{R/j\omega C}{R + (1/j\omega C)} = \frac{R}{j\omega RC + 1}
$$

$$
= \frac{1/C}{j\omega + (1/RC)}. \qquad (7.16)
$$

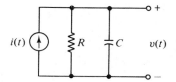

**Figure 7.5**  Circuit of Example 7.2

Thus

$$h(t) = \text{(voltage impulse response)}$$

$$= \frac{1}{C}\epsilon^{-t/RC}u(t). \tag{7.17}$$

From Examples 7.1 and 7.2 it is seen that a voltage impulse in a series *RL* circuit has the effect of immediately establishing an initial current of value $1/L$ in the inductor, regardless of the value of $R$, and a current impulse applied to a parallel *RC* circuit immediately establishes a voltage of value $1/C$ across the capacitor, again regardless of the value of $R$. This is in violation of our previous rules which say that one cannot instantaneously change the current through an inductor or the voltage across a capacitor. These rules, however, were derived only for finite sources, and the impulse function is not a finite source. In the circuit of Figure 7.4 we could add a capacitor in series with the resistor and the initial current would still be just $1/L$, since for instantaneous current changes the inductor is the element that is critical in limiting the current, just as in Figure 7.5 the capacitor is the element that is critical in determining the initial voltage, regardless of how much resistance or inductance is in parallel with it. In both of these examples, the effect of the impulse is simply to establish an initial condition—the circuit response can then be found by solving the initial condition problem, as we did in Chapter 4, with results identical to those found in the examples by use of the Fourier transform. Additional examples are given in the problems at the end of the chapter.

### Pulse and Impulse Response

Let us return to Example 7.1, and this time apply a rectangular pulse of voltage, rather than a true impulse, to the circuit of Figure 7.4. If the pulse has a height $H$ and spans the interval $(0, w)$, it may be expressed in terms of step functions as

$$v(t) = H[u(t) - u(t - w)]. \tag{7.18}$$

Since a unit step voltage $u(t)$ results in the response (Chapter 4)

$$i_s(t) = \frac{1}{R}(1 - \epsilon^{-t/\tau})u(t), \tag{7.19}$$

where $\tau = L/R$, the pulse voltage of equation 7.18 gives a response (by superposition)

$$i(t) = \frac{H}{R}\{(1 - \epsilon^{-t/\tau})u(t) - [1 - \epsilon^{-(t-w)/\tau}]\cdot u(t - w)\}. \qquad (7.20)$$

This response is sketched in Figure 7.6. Let us now increase $H$ and decrease $w$ so that $wH = A$ remains constant. Over the interval $(0, w)$, the current will rise faster but for a shorter interval of time. It will always reach its maximum value at $t = w$, and this value is given by

$$i_{max} = \frac{H}{R}(1 - \epsilon^{-w/\tau}). \qquad (7.21)$$

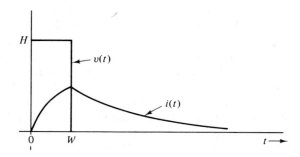

**Figure 7.6**  Pulse Response of Circuit of Figure 7.4

Since $H = A/w$, and we are keeping the area of the pulse ($A$) constant, we may write

$$i_{max} = \frac{A}{Rw}(1 - \epsilon^{-w/\tau}). \qquad (7.22)$$

Now, if we let $w \rightarrow 0$, $i_{max}$ from equation 7.22 is indeterminant, and we must apply l'Hospital's rule. Differentiating numerator and denominator with respect to $w$ gives

$$i_{max} = \lim_{w \to 0} \frac{A}{R\tau} \epsilon^{-w/\tau} = \frac{A}{L}, \qquad (7.23)$$

and since this value now occurs at $t = w = 0$, the total response is simply

$$i(t) = \frac{A}{L} \epsilon^{-t/\tau}u(t), \qquad (7.24)$$

and this is identical to the response of the circuit to the impulse

$$v(t) = A \, \delta(t), \tag{7.25}$$

as may be seen by comparison with equation 7.15. In summary, the response of our circuit to a finite pulse of area $A$ is seen to approach the response to an impulse of area $A$ as the width of the pulse shrinks toward zero. It is also clear that the proper measure of the strength of an impulse is its area, since its effect on the circuit is to establish initial conditions proportional to $A$. [The $A$ in equation 7.25 should not be confused with the height of the impulse, which is, of course, infinite—it is indeed the area of the impulse, since

$$\int_{-\infty}^{+\infty} A \, \delta(t) \, dt = A \int_{-\infty}^{+\infty} \delta(t) \, dt = A.] \tag{7.26}$$

## 7.2 CONVOLUTION

### Convolution Integral

Let us consider a system characterized by an input function $x(t)$, an output $y(t)$, and the system function $H(j\omega)$. Then

$$Y(j\omega) = \mathcal{F}[y(t)] = H(j\omega)X(j\omega)$$
$$= \int_{-\infty}^{+\infty} h(t)\epsilon^{-j\omega t} \, dt \int_{-\infty}^{+\infty} x(t)\epsilon^{-j\omega t} \, dt, \tag{7.27}$$

where $h(t)$ is the impulse response. By changing a variable of integration we may write the product of the two integrals as a double integral:

$$Y(j\omega) = \int_{-\infty}^{+\infty} \int_{-\infty}^{+\infty} h(t)x(\tau)\epsilon^{-j\omega(t+\tau)} \, dt \, d\tau. \tag{7.28}$$

We will now do another change of variable by letting

$$t + \tau = \lambda, \qquad dt = d\lambda,$$

and

$$Y(j\omega) = \int_{-\infty}^{+\infty} \int_{-\infty}^{+\infty} h(\lambda - \tau)x(\tau)\epsilon^{-j\omega\lambda} \, d\lambda \, d\tau. \tag{7.29}$$

We may rewrite this integral as

$$Y(j\omega) = \int_{-\infty}^{+\infty} \left[ \int_{-\infty}^{+\infty} h(\lambda - \tau)x(\tau) \, d\tau \right] \epsilon^{-j\omega\lambda} \, d\lambda$$
$$= \mathcal{F}\left[ \int_{-\infty}^{+\infty} h(\lambda - \tau)x(\tau) \, d\tau \right]. \tag{7.30}$$

Therefore,

$$y(\lambda) = \int_{-\infty}^{+\infty} h(\lambda - \tau)x(\tau)\,d\tau, \tag{7.31}$$

or,

$$y(t) = \int_{-\infty}^{+\infty} h(t - \tau)x(\tau)\,d\tau. \tag{7.32}$$

For physically realizable circuits and systems, the impulse response cannot begin before the impulse occurs; hence

$$h(t) = 0 \qquad (t < 0), \tag{7.33}$$

and

$$h(t - \tau) = 0 \qquad (\tau > t). \tag{7.34}$$

Thus, for physically realizable circuits,

$$\boxed{y(t) = \int_{-\infty}^{t} h(t - \tau)x(\tau)\,d\tau.} \tag{7.35}$$

This is called a *convolution integral,* and by another change of variable it may be written in an alternative form: if we let

$$t - \tau = \lambda, \quad \text{and} \quad d\tau = -d\lambda,$$

then

$$\tau \to -\infty \Rightarrow \lambda \to +\infty, \quad \text{and} \quad \tau = t \Rightarrow \lambda = 0.$$

Hence

$$y(t) = \int_{0}^{\infty} h(\lambda)x(t - \lambda)\,d\lambda,$$

or

$$\boxed{y(t) = \int_{0}^{\infty} h(\tau)x(t - \tau)\,d\tau.} \tag{7.36}$$

Both equations 7.35 and 7.36 involve convolution integrals, and we see that this convolution operation is the time-domain counterpart to the multiplication of Fourier transforms in the frequency domain (equation 7.27). The convolution operation, as represented by either equation 7.35 or 7.36, is usually denoted as

$$h(t) * x(t) = y(t), \tag{7.37}$$

and this corresponds to the operation

$$H(j\omega)X(j\omega) = Y(j\omega) \tag{7.38}$$

in the frequency domain. We will now discuss some physical interpretations of this convolution operation.

### Superposition Interpretation of Convolution

Let us examine the convolution integral in the form given in equation 7.35:

$$y(t) = h(t)*x(t) = \int_{-\infty}^{t} h(t-\tau)x(\tau)\,d\tau. \qquad (7.39)$$

In Figure 7.7 are sketched a physically realizable impulse response $h(t)$ and an input function $x(t)$. Also sketched are $h(\tau - t)$ and $h(t - \tau)$ as functions of the variable $\tau$. Part (e) of the figure shows the integrated product of $h(t - \tau)$ and $x(\tau)$—that is, at any instant of time $t$, $y(t)$ is the area under the product curve, indicated by the crosshatching in the figure. As time goes on, one can visualize the curve $h(t - \tau)$ "sliding past" $x(\tau)$, and at each time

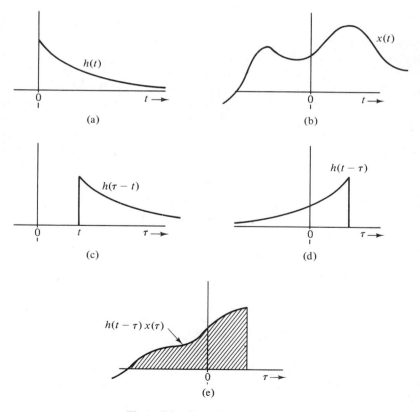

**Figure 7.7** Convolution Operation

epoch the output $y(t)$ is given by the integrated product of the two curves. If the convolution integral is written in the alternative form of equation 7.36, then the roles of $h(t)$ and $x(t)$ are reversed, and it is $x(t)$ which is turned around in time and translated. In either case, however, the convolution operation is seen to consist of four basic steps:

1. A time reversal.
2. A translation.
3. A multiplication.
4. An integration.

Suppose now that we approximate our forcing function $x(\tau)$ by a sum of rectangular pulses, as indicated in Figure 7.8. If $\Delta\tau$ is sufficiently small, we may approximate each pulse by an impulse of area $x(\tau)\,\Delta\tau$, just as we did in the previous section. [Of course, in the limit, as $\Delta\tau \longrightarrow 0$, each impulse has an infinitesimal area, since $x(\tau)\,\Delta\tau \longrightarrow 0$, and the spacing between them becomes infinitesimal.] We may visualize each narrow pulse at time $\tau_i$ "triggering" an impulse response starting at $\tau_i$, with an amplitude proportional to $x(\tau_i)$. The output $y(t)$ is then just the sum at time $t$ of the "tails" of all these impulse responses, as indicated in Figure 7.8(a). However, a little thought will make it clear that exactly the same result will be obtained if we start with a single unit impulse response, turned around and shifted as in part (b) of the figure, and then weight the tail of this impulse response with the heights of the various pulse components of $x(\tau)$, and sum. The operations indicated by parts (a) and (b) of the figure are identical—only the mathematical implementation is different. Thus the convolution integral is really

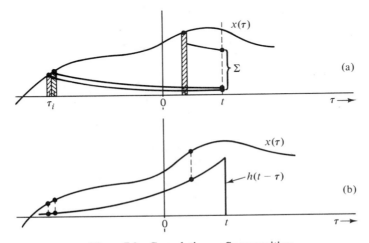

**Figure 7.8** Convolution as Superposition

just a consequence of the linearity of the system and the validity of super-position. We have tried to indicate this here by the use of a rather heuristic argument; the convolution integral can, however, be derived in rigorous fashion as the limit of a sum, using superposition.

## 7.3 DETERMINATION OF NETWORK IMPULSE RESPONSE

The convolution integral makes it possible to immediately write down an integral expression for the output function of a network for any given forcing function, provided that the impulse response is known. This gives the impulse a certain preeminent position among forcing functions and makes it desirable to be able to determine impulse responses of networks. These may be determined by either calculation or measurement. If, as we have assumed throughout this text, an adequate model of the network involving our linear circuit elements is available, then we may calculate the impulse response mathematically. Sometimes, however, a model for a network is not known—the network is, so to speak, in a "black box," with only an input and output available. If we know that what is in the black box is linear, however, then the impulse response tells us as much about the input–output behavior of the network (through the convolution integral) as does a detailed network model. In this case the impulse response cannot be calculated, but it can be physically measured. Unfortunately, the impulse function is not easy to physically approximate. Its integral, the step function, is easy to physically generate, however; an excellent approximation to a step function of voltage or current may be obtained by the simple operation of opening or closing a switch, provided that the sources are of sufficient capacity to act essentially as ideal sources. Hence, since the network is linear, we may obtain a good approximation of the impulse response of our black box by using a unit step function as a test input, and then differentiating the output by performing a graphical or numerical differentiation. (For a computer simulation, a digital representation of the impulse response is required anyway.)

If the network model is available, then of course the impulse response can be calculated. This can be done, as we have seen, by finding the inverse Fourier transform of the system function. (We will discuss the transform method of finding the impulse response in much more detail in Chapter 8, when we discuss Laplace transforms.) In some instances, as in Examples 7.1 and 7.2, it is easier to consider the problem as an initial-condition problem, keeping in mind that a unit impulse of voltage applied in series with an induc-tor instantaneously establishes unit flux linkages in the inductor, and a unit impulse of current applied to a capacitor instantaneously establishes unit charge on the capacitor. In other cases, it is easier to reason in terms of an applied step function, and differentiate the result, just as we would do if we were trying to physically measure the impulse response. This is illustrated in the following example.

### Example 7.3

We will calculate the impulse response of the circuit of Figure 7.9 by first writing the step response and then differentiating. From the discussions in Chapter 4 we know that the time constant of the circuit is

$$\tau = \frac{R_1 R_2}{R_1 + R_2} C. \tag{7.40}$$

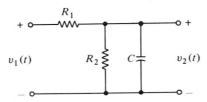

**Figure 7.9** Circuit of Example 7.3

If $v_1(t) = u(t)$, then $v_2(0) = 0$ and $v_2(\infty) = R_2/(R_1 + R_2)$. Hence the unit step response is

$$v_2(t)_{\text{step}} = \frac{R_2}{R_1 + R_2} (1 - \epsilon^{-t/\tau}) u(t). \tag{7.41}$$

By differentiating this unit step response we obtain the unit impulse response:

$$h(t) = \frac{1}{R_1 C} \epsilon^{-t/\tau} u(t). \tag{7.42}$$

### Impulse Response in Terms of Homogeneous Equation Solutions

Let us suppose that we have used Kirchhoff's laws to write a set of simultaneous network differential equations, and that we have reduced this set of equations to a single differential equation (using, for example, differential operator notation) relating the input and output variables. As an example, consider the second-order equation

$$(a_2 D^2 + a_1 D + a_0)y(t) = (b_1 D + b_0)x(t), \tag{7.43}$$

where $x(t)$ is the input and $y(t)$ the output variable. If $h(t)$ is the impulse response, we know from the convolution integral that

$$y(t) = \int_{-\infty}^{t} h(t - \tau)x(\tau)\, d\tau. \tag{7.44}$$

Let us now differentiate $y(t)$ twice:

$$y'(t) = h(t - \tau)x(\tau)\Big|_{\tau=t} + \int_{-\infty}^{t} h'(t - \tau)x(\tau)\,d\tau$$

$$= h(0)x(t) + \int_{-\infty}^{t} h'(t - \tau)x(\tau)\,d\tau \tag{7.45}$$

$$y''(t) = h(0)x'(t) + h'(t - \tau)x(\tau)\Big|_{\tau=t}$$

$$+ \int_{-\infty}^{t} h''(t - \tau)x(\tau)\,d\tau$$

$$= h(0)x'(t) + h'(0)x(t) + \int_{-\infty}^{t} h''(t - \tau)x(\tau)\,d\tau. \tag{7.46}$$

Substituting these quantities into equation 7.43 yields

$$a_2 h(0)x'(t) + a_2 h'(0)x(t) + a_1 h(0)x(t)$$

$$+ \int_{-\infty}^{t} x(\tau)[a_2 h''(t - \tau) + a_1 h'(t - \tau) + a_0 h(t - \tau)]\,d\tau$$

$$= b_1 x'(t) + b_0 x(t). \tag{7.47}$$

If we now choose $h(0)$ and $h'(0)$ such that

$$a_2 h(0) = b_1 \tag{7.48}$$

$$a_2 h'(0) + a_1 h(0) = b_0, \tag{7.49}$$

then the integral in equation 7.47 must equal zero. For this to be true for arbitrary $x(\tau)$ requires that

$$a_2 h''(t - \tau) + a_1 h'(t - \tau) + a_0 h(t - \tau) = 0, \tag{7.50}$$

or

$$a_2 h''(t) + a_1 h(t) + a_0 h(t) = 0, \tag{7.51}$$

which is just the homogeneous equation associated with equation 7.34. We know that this homogeneous equation has two linearly independent solutions, or a general solution of the form

$$h(t) = C_1 \epsilon^{+s_1 t} + C_2 \epsilon^{+s_2 t}, \tag{7.52}$$

where $s_1$ and $s_2$ are the roots of the characterictic equation

$$a_2 s^2 + a_1 s + a_0 = 0. \tag{7.53}$$

Hence we may find our impulse response if we know the solutions of the homogeneous equation—in this instance, $h(t)$ is given by equation 7.52, where $s_1$ and $s_2$ are found from equation 7.53 and the constants $C_1$ and $C_2$ are evaluated using the initial conditions $h(0)$ and $h'(0)$ found from equations 7.48 and 7.49. We will illustrate this method with an example.

### Example 7.4

Let us find the impulse response associated with the equation

$$(D^2 + 3D + 2)y(t) = x(t). \tag{7.54}$$

The characteristic equation for this example is

$$S^2 + 3S + 2 = 0, \tag{7.55}$$

which means that the impulse response has the general form

$$h(t) = C_1 \epsilon^{-t} + C_2 \epsilon^{-2t}. \tag{7.56}$$

By following the procedure outlined in equations 7.46 through 7.49, we see that

$$h(0) = 0 \tag{7.57}$$

and

$$h'(0) = 1$$

(which may be easily seen by setting $b_1 = 0$, $b_0 = 1$, and $a_2 = 1$). Applying these initial conditions to equation 7.56 gives

$$C_1 + C_2 = 0 \tag{7.58}$$
$$-C_1 - 2C_2 = 1.$$

Solving these equations gives $C_1 = 1$ and $C_2 = -1$, so

$$h(t) = \epsilon^{-t} - \epsilon^{-2t} \qquad (t \geq 0). \tag{7.59}$$

Although we have used a second-order example to illustrate the relationship between the impulse response and the homogeneous equation solutions, the procedure indicated in equations 7.45 through 7.56 may be used on an equation of any form. In fact, if the equation is of the form

$$(D^n + a_{n-1}D^{n-1} + \ldots + a_0)y(t) = x(t), \tag{7.60}$$

then applying this procedure leads to the conclusion that $h(t)$ is the solution to the homogeneous equation with

$$h(0) = h'(0) = h''(0) = \ldots = h^{(n-2)} = 0$$
$$h^{(n-1)} = 1. \tag{7.61}$$

(Example 7.4 illustrates this in the second-order case.)

In summary, we see that it is possible, by means of the impulse response and convolution integral, to write down a time domain expression for the output function of a network for an arbitrary forcing function input, in terms of the solutions of the homogeneous equation of the network. These ideas will be expanded upon when we discuss network state variables in Chapter 9.

## 7.4 DIGITAL COMPUTER SIMULATION OF ANALOG NETWORKS

The various analysis techniques that we have been discussing can be applied to very simple circuits, to networks of great complexity, or to complex systems that may involve electrical networks in conjunction with mechanical, thermal, or other types of physical elements. Often, in the design of a system (such as a control system for an aircraft or space vehicle, for example) it is necessary to evaluate the system performance over a wide variety of input or disturbance signals. If the system is a complex one, it is usually not feasible to do such an evaluation analytically; in fact, very often, as in the case of random disturbance signals, an exact mathematical expression for the system inputs will not be available, and a computer simulation will be required. In fact, most modern systems require extensive computer simulation studies before a final design is completed and approved. A detailed study of computer simulation methods (analog and digital) is beyond the scope of this text, but we will discuss one small aspect of the subject which is closely related to the other material in this chapter—a simple digital computer implementation of the convolution operation.

### Digital Simulation of the Network Input–Output Relationship

Suppose that we have a linear network or filter which we want to simulate on a digital computer, insofar as its input–output relationship is concerned. (The network may, of course, be just one small part of a large-scale system.) We will assume that we are able to write the equations of the network and to reduce these to a single differential equation (e.g., equation 7.43) that relates the input and output variables of the network.

In order to digitally solve or simulate our differential equation, we must quantize both the amplitude and time variables. With the precision available in the modern digital computer, the amplitude quantization really presents no problems, either practically or theoretically. The time quantization, however, does change the problem theoretically, by converting our differential equation to what is called a *difference equation*, and we will discuss the relationship between differential and difference equations in some detail in Chapter 9. Since the present chapter is concerned with things related to impulse response and convolution, we will look at digital simulation here from the point of view of numerically evaluating the convolution integral.

Let us suppose that we have available a set of numbers representing the values of the impulse response of our system at time points separated by an interval of $\Delta t$ seconds, and a similar set of sample values of the input function. These sample values could have been made available to us by specification, measurement, or computation from a continuous-time model. Using these sample values, we may construct various approximations to the original continuous functions by using different sorts of interpolations between the samples. For example, we might construct a piecewise constant approximation of the original functions by using a constant interpolation between samples. The numerical procedure for this type of approximation is quite simple (see Problem 7.11). A much better approximation, however, is obtained by using a straight-line approximation between sample values, as one does in evaluating integrals using the trapezoidal formula. Let us investigate the numerical evaluation of the convolution integral using the straight-line approximation.

We will first calculate the integrated product of two straight-line segments spanning the same time interval, as indicated in Figure 7.10. Since the integrated product of these segments does not depend upon the position of the interval along the time axis, we may simplify the calculation by translating the segments back to the origin, as in part (b) of the figure. We then have

$$x(t) = at + b$$
$$h(t) = ct + d \qquad (0 \le t \le \Delta t) \qquad (7.62)$$

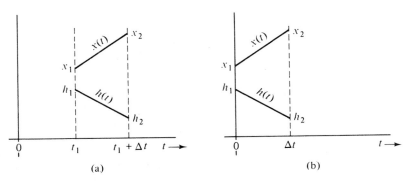

$$(a) \qquad\qquad\qquad\qquad (b)$$

**Figure 7.10** Two Line Segments Spanning the Same Time Interval

with

$$a = \frac{x_2 - x_1}{\Delta t}, \qquad b = x_1$$

$$c = \frac{h_2 - h_1}{\Delta t}, \qquad d = h_1. \tag{7.63}$$

Integrating the product, we obtain

$$I = \int_0^{\Delta t} (at + b)(ct + d)\, dt = \int_0^{\Delta t} [act^2 + (bc + ad)t + bd]\, dt$$
$$= \tfrac{1}{3} ac(\Delta t)^3 + \tfrac{1}{2}(bc + ad)(\Delta t)^2 + bd(\Delta t). \tag{7.64}$$

Substituting the values from equations 7.63 yields

$$I = \frac{\Delta t}{3}\left[x_1 h_1 + \frac{1}{2}(x_1 h_2 + x_2 h_1) + x_2 h_2\right]. \tag{7.65}$$

To evaluate the convolution integral, we need to express the sum of a number of such integrated products of overlapping line segments. Let us consider an input signal that starts at $t = 0$, and we will use the form of the convolution integral in which the impulse response is turned around in time and translated, as indicated in Figure 7.11. From the result of equation 7.65,

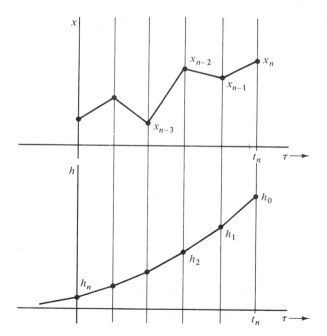

**Figure 7.11**  Numerical Convolution with Straight-line Interpolation

it is easily seen that the complete convolution integral will be approximated by

$$y_n = \frac{\Delta t}{3}\left[ x_{n-1}h_1 + \frac{1}{2}(x_{n-1}h_0 + x_nh_1) + x_nh_0 \right]$$

$$+ \left[ x_{n-2}h_2 + \frac{1}{2}(x_{n-2}h_1 + x_{n-1}h_2) + x_{n-1}h_1 \right]$$

$$+ \cdots$$

$$+ \left[ x_0h_n + \frac{1}{2}(x_0h_{n-1} + x_1h_n) + x_1h_{n-1} \right],$$

or

$$y_n = \frac{\Delta t}{3}\sum_{m=1}^{n} x_{n-m}h_m + \frac{1}{2}(x_{n-m}h_{m-1} + x_{n-m+1}h_m) + x_{n-m+1}h_{m-1}. \qquad (7.66)$$

Equation 7.66 may be easily programmed on any digital computer and the $y_n$ (output) sequence calculated in real time. Of course, as $n$ increases, the length of the calculation required to compute $y_n$ increases. In practice, however, most impulse responses die out in time—i.e., after a certain length of time (sometimes called the *settling time of the system*) there will be negligible energy left in the remaining tail of the impulse response. In this case, for some sampling time $k$, we may set

$$h_m = 0 \qquad \text{for } m \geq k,$$

or

$$h_{m-1} = 0 \qquad \text{for } m > k.$$

Then equation 7.66 may be written as

$$y_n = \begin{cases} \dfrac{\Delta t}{3}\displaystyle\sum_{m=1}^{n} \text{(etc.)} & \text{for } n \leq k \\[2ex] \dfrac{\Delta t}{3}\displaystyle\sum_{m=1}^{k} \text{(etc.)} & \text{for } n > k. \end{cases} \qquad (7.67)$$

Thus, after $k$ steps, the length of the calculation remains constant. With a modern high-speed computer, the calculation of equation 7.66 is extremely rapid, even with hundreds of sample values, and hence our linear input–output network relationship may be simulated with great accuracy. A similar, but iterative type of simulation will be discussed in Chapter 9 in connection with state-variable network models.

We have illustrated one way of digitally simulating the input–output relationship of a linear circuit or system, using sample values of the impulse response. This is one of the best practical methods available, if the samples of the impulse response are known. It should be apparent to the reader at this point that a numerical evaluation of the convolution integral is possible for any specified type of numerical -signal representation. We might, for example, expand our input signal and impulse response in any series of orthogonal functions and derive an expression for the convolution integral in terms of the expansion coefficients. For most practical applications, however, the straight-line-segment approximation using the signal sample values is very convenient and accurate, and the degree of error of the approximation for a given sampling rate is fairly evident, if the general nature of the waveforms is known. (There is another type of interpolating function, of the form $\sin t/t$, which is often associated with sample values of band-limited functions. The use of these functions arises from the *Nyquist sampling theorem*, which we will not discuss here. For a detailed discussion of the sampling theorem, the reader may refer to a variety of texts that treat signal analysis in more depth.)[2]

## PROBLEMS

**7.1**  Show that, as $a \longrightarrow \infty$, the function

$$f(t) = \frac{a}{\sqrt{\pi}} \epsilon^{-a^2 t^2}$$

approaches a unit impulse function.

**7.2**  Find the value of $K$ such that

$$f(t) = K\epsilon^{-t/a}u(t)$$

approaches a unit impulse function as $a \longrightarrow 0$.

**7.3**  Evaluate the following integrals:

(a)  $\displaystyle\int_{-\infty}^{+\infty} (\sin x)\,\delta(x - \pi)\,dx.$

(b)  $\displaystyle\int_{-\infty}^{+\infty} \epsilon^{j\omega t}\,\delta(t - \tau)\,dt.$

(c)  $\displaystyle\int_{1}^{\infty} \delta(t)\cos \omega t\,dt.$

(d)  $\displaystyle\int_{-1}^{+1} \delta(t)\epsilon^{-t}\sin \omega t\,dt.$

(e)  $\displaystyle\int_{-\infty}^{+\infty} f(\tau)\,\delta(t - \tau)\,d\tau.$

[2]See, for example, R. J. Schwarz and B. Friedland, *Linear Systems*, McGraw-Hill Book Company, New York, 1965, or R. A. Gabel and R. A. Roberts, *Signals and Linear Systems*, John Wiley & Sons, New York, 1973.

**7.4** For the circuits of Figure 7.12, find the impulse responses for the input and output quantities indicated, by first evaluating the unit step responses and then differentiating.

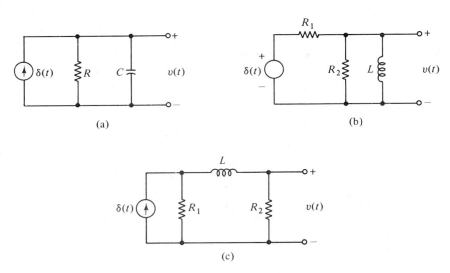

(a)

(b)

(c)

**Figure 7.12** Circuits for Problem 7.4

**7.5** Find the impulse responses associated with the following equations, in terms of the solutions of the related homogeneous equations.
(a) $\ddot{y}(t) + 5\dot{y}(t) + 6y(t) = 2\dot{x}(t)$.
(b) $\ddot{y}(t) + 2\dot{y}(t) + y(t) = x(t)$.
(c) $\ddot{y}(t) + 4y(t) = \dot{x}(t) + 2x(t)$.

**7.6** Find the impulse response for the current in the circuit of Figure 7.13.

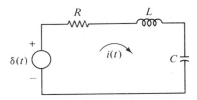

**Figure 7.13** Circuit for Problem 7.6

**7.7** Sketch the result of convolving the pairs of functions indicated in Figure 7.14. Show enough information on the sketches to completely specify the resulting function.

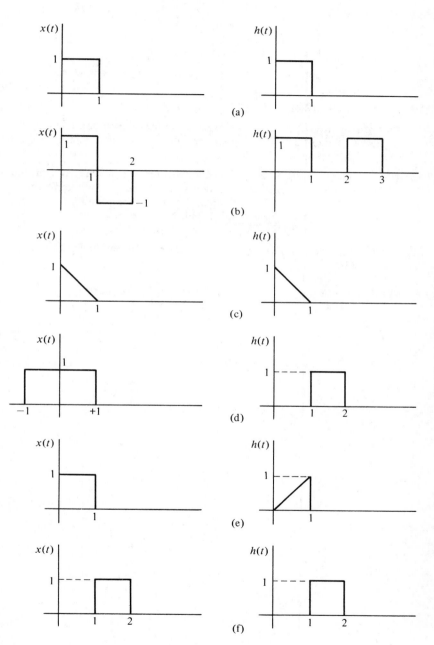

**Figure 7.14** Waveforms for Problem 7.7

**7.8** Find $x(t) * h(t)$, if
    (a) $x(t) = \epsilon^{-t}u(t)$,     $h(t) = \epsilon^{-at}u(t)$.
    (b) $x(t) = \sin t$,     $h(t) = \epsilon^{-at}u(t)$.
    (c) $x(t) = (\sin t)u(t)$,     $h(t) = u(t) - u(t - \tau)$.

**7.9** Show that convolution with a unit step function corresponds to a simple integration.

**7.10** Show that $x(t) * \delta(t) = x(t)$. What is $x(t) * \delta(t - \tau)$?

**7.11** Derive a numerical procedure for evaluating the convolution integral, using a piecewise constant (stair-step) approximation of the impulse response and input function.

**7.12** (Programming problem.) Program a digital computer to find the output sample value sequence of a circuit, using a straight-line interpolation (equation 7.66 or 7.67). Choose a fairly simple impulse response and input function (such as exponentials, for example), find their sample values, and use these to test the correctness and accuracy of your program.

**7.13** If $v(t) = \epsilon^{-kt}u(t)$ is the voltage across a 1-$\Omega$ resistor, find the total energy in the pulse by
    (a) Direct integration.
    (b) Applying Parseval's theorem and the Fourier transform.

# 8

# Laplace Transforms
# and Complex Frequency

In Chapter 6 we considered one example of a "transform method"; in particular, we used the Fourier integral to establish a one-to-one correspondence between time functions and functions of the frequency variable $\omega$. The Fourier integral operator is, however, only one example of what are usually called *integral transforms*. It is perhaps the most important example, because of its great theoretical usefulness in many areas of physics, signal analysis, probability theory, information theory, and other fields. In fact, Fourier series and transforms constitute one of the primary mathematical tools for analyzing physical phenomena. Also, since Fourier series and transforms represent basically the resolution of waveforms into sinusoidal components, they provide intuitive insight into problems of system behavior, by relating many aspects of this behavior to frequency response.

The Laplace transform is another integral transform but somewhat better suited to the particular problem of finding transient responses of linear networks and systems. Some of the advantages of the Laplace transform in this regard are:

1. The Laplace transform may be defined for certain very practical functions for which the Fourier transform does not exist in the usual sense. Consider, for example, the unit step function or a sinusoid defined over the interval $(0, \infty)$. Neither of these functions is absolutely integrable, and to Fourier-analyze them requires some extension of the Fourier integral theory presented in Chapter 6. Both functions are, however, very important from an engineering point of view.
2. When using the Laplace transform, initial conditions are accounted

for as soon as the equations are transformed, and the response to initial conditions and the response due to the forcing function are obtained together, as parts of the total response. (We recall that in using the Fourier transform to find network transient responses, a separate solution of the homogeneous equation is required to find the part of the solution due to nonzero initial conditions, if any.)

3. Linear system theory is greatly enriched by the use of the complex frequency plane and pole–zero analysis, which are closely related to Laplace transforms.

4. Notation and algebraic manipulation of transform equations and transfer functions is somewhat simplified by using Laplace rather than Fourier transforms.

## 8.1 THE LAPLACE TRANSFORM

### Definition and Relationship to Fourier Transforms

Throughout this chapter we shall restrict our consideration to time functions that are zero for negative time. Physically speaking, this is not too great a restriction, since all networks or systems must be actuated at some time epoch, which we may consider to be $t = 0$, even if, as in the case of a power-generating system, the system is designed to operate in the steady state. If we only desire the steady-state part of the solution, i.e., the part that is nonzero after the transient has died out, we can, after all, simply discard the transient part of the solution. Examples of time functions that are zero for negative time are the unit step function $u(t)$, $\epsilon^{-\alpha t}u(t)$, $(\cos \omega t)u(t)$, and so forth. Of these three functions mentioned, $\epsilon^{-\alpha t}u(t)$ is absolutely integrable provided that $\alpha > 0$, but the other two functions are not absolutely integrable even though they are zero for negative time, and hence do not possess Fourier transforms in the usual sense. Suppose, however, that we define a new transform by first multiplying our time function by a *convergence factor* $\epsilon^{-\sigma t}$, and then taking the Fourier transform of this product. This new transform is called the *Laplace transform*, and we will designate it by the operator $\mathcal{L}$, to distinguish it from the Fourier transform operator $\mathcal{F}$. Since our time function is assumed to be zero for negative time, we may write the Laplace transform as

$$\mathcal{L}[f(t)] \triangleq \mathcal{F}[f(t)\epsilon^{-\sigma t}]$$
$$= \int_{-\infty}^{+\infty} f(t)\epsilon^{-\sigma t}\epsilon^{-j\omega t}\, dt = \int_{0}^{\infty} f(t)\epsilon^{-(\sigma+j\omega)t}\, dt. \qquad (8.1)$$

The exponent factor $(\sigma + j\omega)$ is called a *complex frequency*, and is usually denoted by the complex variable $s$, just as in previous discussions we have denoted complex exponentials by $\epsilon^{st}$. Thus the Laplace transform is usually indicated by

$$\mathcal{L}[f(t)] = F(s) = \int_0^\infty f(t)\epsilon^{-st}\, dt, \tag{8.2}$$

and is a function of the complex variable $s$. Since

$$f(t)\epsilon^{-\sigma t} = \mathcal{F}^{-1}[F(s)]$$

$$= \frac{1}{2\pi} \int_{-\infty}^{+\infty} F(s)\epsilon^{j\omega t}\, d\omega, \tag{8.3}$$

multiplying by $\epsilon^{\sigma t}$ gives

$$f(t) = \frac{1}{2\pi} \int_{-\infty}^{+\infty} F(s)\epsilon^{(\sigma + j\omega)t}\, d\omega. \tag{8.4}$$

Since $s = \sigma + j\omega$, $ds = j\, d\omega$, and changing the variable of integration to $s$ gives

$$f(t) = \frac{1}{2\pi j} \int_{\sigma - j\infty}^{\sigma + j\infty} F(s)\epsilon^{st}\, ds$$

$$= \mathcal{L}^{-1}[F(s)]. \tag{8.5}$$

The $f(t)$ and $F(s)$ as given by equations 8.2 and 8.5 constitute a *Laplace transform pair*, or, more particularly, a *single-sided Laplace transform pair*. (Double-sided Laplace transforms, which may be applied to certain functions that are not zero for negative time, will not be treated here.)

### Existence and Uniqueness of Laplace Transforms

The Laplace transform of a function $f(t)$ is said to exist if the integral of equation 8.2 converges (i.e., approaches a finite limit as the upper limit on the integral approaches infinity) for some range of the variable $s$. We know that the Fourier integral exists if a function is absolutely integrable, and the Laplace transform is the Fourier transform of $f(t)\epsilon^{-\sigma t}$. Hence the Laplace transform will exist if

$$\int_0^\infty |f(t)\epsilon^{-\sigma t}|\, dt = \int_0^\infty |f(t)|\, \epsilon^{-\sigma t}\, dt < \infty. \tag{8.6}$$

Functions for which this integral is finite, for some value of $\sigma$, are said to be functions of *exponential order*. If $\sigma$ is a large positive number, then $\epsilon^{-\sigma t}$ is a powerful convergence factor, which, when multiplied by $f(t)$, will cause the product to approach zero for large $t$ even for many functions $f(t)$ which themselves grow with time. Examples of functions of exponential order (and the corresponding range of $\sigma$) include

| $f(t)$ | Range of $\sigma$ |
|:---:|:---:|
| $u(t)$ | $\sigma > 0$ |
| $t^n u(t)$ | $\sigma > 0$ |
| $\epsilon^{-\alpha t} \cos(\omega t + \theta)u(t)$ | $\sigma > -\alpha$ |
| $\epsilon^{\alpha t} u(t)$ | $\sigma > \alpha$ |

Thus, even a growing exponential (for positive time) possesses a Laplace transform, although some functions grow too fast to be killed off by the exponential convergence factor. An example is the function $f(t) = \epsilon^{t^2}$. However, such functions almost never appear in linear system analyses. The minimum value of $\sigma$ for which the integral of equation 8.6 exists is called the *abscissa of absolute convergence* associated with $f(t)$. It is important to know the abscissa of absolute convergence in using the inverse transform integral of equation 8.5. In this treatment we shall not be evaluating our inverse transforms in this way, since integrations over the complex plane go somewhat beyond the prerequisite mathematical background assumed for this text, and hence we will not specifically indicate the abscissa of absolute convergence each time that we write a Laplace transform pair, and we will always assume that our functions $f(t)$ possess Laplace transforms.

As in the case of Fourier transforms, it may be shown that Laplace transforms are unique, except for functions that differ from each other at discrete points, or a set of points of measure zero. For practical purposes we need not consider this exception, and may simply assume that if a function possesses a Laplace transform, then that transform is unique. This unique-

**Table 8.1**  Selected Laplace Transforms

| | $f(t)$ | $F(s)$ |
|:---|:---:|:---:|
| 1. | $u(t)$ | $\dfrac{1}{s}$ |
| 2. | $\epsilon^{-\alpha t}u(t)$ | $\dfrac{1}{s + \alpha}$ |
| 3. | $(\cos \omega_0 t)u(t)$ | $\dfrac{s}{s^2 + \omega_0^2}$ |
| 4. | $(\sin \omega_0 t)u(t)$ | $\dfrac{\omega_0}{s^2 + \omega_0^2}$ |
| 5. | $t^n u(t)$ | $\dfrac{n!}{s^{n+1}}$ |
| 6. | $\epsilon^{-\alpha t}(\cos \omega_0 t)u(t)$ | $\dfrac{s + \alpha}{(s + \alpha)^2 + \omega_0^2}$ |
| 7. | $\epsilon^{-\alpha t}(\sin \omega_0 t)u(t)$ | $\dfrac{\omega_0}{(s + \alpha)^2 + \omega_0^2}$ |
| 8. | $t^n \epsilon^{-\alpha t}u(t)$ | $\dfrac{n!}{(s + \alpha)^{n+1}}$ |

ness property means that we can build up a table of Laplace transform pairs, using the relatively easily evaluated integral of equation 8.2, and then find inverse transforms from the table, without the necessity of each time evaluating the difficult "contour" integration of equation 8.5.

### Short Table of Laplace Transforms

Extensive tables of both Fourier and Laplace transforms have been published, but for ordinary linear network transient analysis, knowledge of only a few transforms will usually suffice. The reason for this will become evident shortly when we consider partial fraction expansions. Table 8.1 lists some of the most commonly used Laplace transforms. In the problems at the end of the chapter the reader is asked to derive several of these transforms. We shall derive number 2 (one of the most important ones) as an example.

### Example 8.1

The Laplace transform of $\epsilon^{-\alpha t}u(t)$ is given by

$$F(s) = \int_0^\infty \epsilon^{-\alpha t}\epsilon^{-st}\,dt = \int_0^\infty \epsilon^{-(s+\alpha)t}\,dt$$

$$= -\frac{1}{s+\alpha}[\epsilon^{-(s+\alpha)t}]_0^\infty$$

$$= \frac{1}{s+\alpha}[\epsilon^{-(\sigma+\alpha+j\omega)t}]_\infty^0$$

$$= \frac{1}{s+\alpha}[\epsilon^{-(\sigma+\alpha)t}(\cos\omega t - j\sin\omega t)]_\infty^0$$

$$= \frac{1}{s+\alpha} \qquad \text{for } \sigma > -\alpha. \tag{8.7}$$

Here the Laplace integral converges provided $\sigma > -\alpha$, $-\alpha$ being the abscissa of absolute convergence.

### Properties of Laplace Transforms

We shall now examine some of the properties of Laplace transforms upon which their usefulness in solving differential equations depends. These properties are quite similar to the Fourier transform properties discussed in Chapter 6, and some of the derivations are left for the problems at the end of the chapter.

1. *Linearity*

$$\mathcal{L}[af_1(t) + bf_2(t) + \ldots] = aF_1(s) + bF_2(s) + \ldots. \tag{8.8}$$

(See Problem 8.3.)

2. *Transforms of derivatives*

$$\mathcal{L}\left[\frac{d}{dt}\,f(t)\right] = sF(s) - f(0). \tag{8.9}$$

*Proof:*

$$\mathcal{L}\left[\frac{d}{dt}\,f(t)\right] = \int_0^\infty \frac{d}{dt}\,f(t)\epsilon^{-st}\,dt. \tag{8.10}$$

We may integrate this equation by parts, letting

$$u = \epsilon^{-st} \quad \text{and} \quad dv = \frac{d}{dt}\,f(t)\,dt = df(t)$$

in the integration-by-parts formula

$$\int_a^b u\,dv = uv\Big|_a^b - \int_a^b v\,du. \tag{8.11}$$

Since $du = -s\epsilon^{-st}\,dt$ and $v = f(t)$, we obtain

$$\mathcal{L}\left[\frac{d}{dt}\,f(t)\right] = f(t)\epsilon^{-st}\Big|_0^\infty + \int_0^\infty sf(t)\epsilon^{-st}\,dt. \tag{8.12}$$

Let us now examine the first term on the right-hand side of this equation. As $t \to 0$, $\epsilon^{-st} \to 1$, and the entire term approaches $f(0)$. [Since we let $t \to 0$ from the right, we should really write $f(0+)$, which may differ from $f(0-)$ if $f(t)$ is discontinuous at the origin. Using the value $f(0+)$ implies that we must use the initial condition on $f(t)$ evaluated just after a switching action or discontinuity in the network occurs. Hereafter, we will just write $f(0)$, with the assumption that it means $f(0+)$.] Now let us consider

$$\lim_{t\to\infty} f(t)\epsilon^{-st} = \lim_{t\to\infty} f(t)\epsilon^{-\sigma t}\epsilon^{-j\omega t}$$

$$= \lim_{t\to\infty} f(t)\epsilon^{-\sigma t}[\cos \omega t - j \sin \omega t]. \tag{8.13}$$

This approaches zero provided that $f(t)$ is killed off by $\epsilon^{-\sigma t}$ as $t \to \infty$; in other words, if $f(t)$ is of exponential order and $\sigma$ is greater than the abscissa of absolute convergence. Since we are assuming these conditions for the existence of the Laplace transform, equation 8.12 becomes

$$\mathcal{L}\left[\frac{d}{dt}\,f(t)\right] = -f(0) + s \int_0^\infty f(t)\epsilon^{-st}\,dt$$

$$= sF(s) - f(0). \tag{8.14}$$

To evaluate $\mathcal{L}\left[\dfrac{d^2}{dt^2}f(t)\right]$, we may simply repeat the use of formula 8.9; i.e.,

$$\mathcal{L}\left[\frac{d^2}{dt^2}f(t)\right] = s[sF(s) - f(0)] - \dot{f}(0), \tag{8.15}$$

where $\dot{f}(0) \triangleq df/dt\,|_{0+}$. Hence

$$\mathcal{L}\left[\frac{d^2}{dt^2}f(t)\right] = s^2F(s) - sf(0) - \dot{f}(0). \tag{8.16}$$

Similarly, we may evaluate the transform of the third derivative:

$$\mathcal{L}\left[\frac{d^3}{dt^3}f(t)\right] = s[s^2F(s) - sf(0) - \dot{f}(0)] - \ddot{f}(0)$$
$$= s^3F(s) - s^2f(0) - s\dot{f}(0) - \ddot{f}(0). \tag{8.17}$$

Note that if all the initial values are zero, and $s = j\omega$, then the differentiation formulas are identical to those for the Fourier transform. The fact that initial values are accounted for explicitly in using the Laplace transform is one of its most valuable features for the solution of transient problems. (Of course, the basic reason that these initial conditions appear is that we have restricted ourselves to time functions which are zero for negative time— had we defined the Fourier integral with a lower limit of zero, the initial conditions would also have appeared in the differentiation formulas for the Fourier integral.)

### 3. *Transforms of integrals*

$$\mathcal{L}\left[\int_{-\infty}^{t} f(t)\,dt\right] = \frac{F(s)}{s} + \frac{\int_0}{s}, \tag{8.18}$$

where the symbol $\int_0$ denotes the value of the integral at $t = 0$; i.e.,

$$\int_0 \triangleq \int_{-\infty}^{0} f(t)\,dt. \tag{8.19}$$

*Proof*: Since

$$f(t) = \frac{d}{dt}\int_{-\infty}^{t} f(t)\,dt, \tag{8.20}$$

and

$$\mathcal{L}[f(t)] = F(s), \tag{8.21}$$

$$\mathcal{L}\left[\frac{d}{dt}\int_{-\infty}^{t} f(t)\,dt\right] = F(s) = s\mathcal{L}\left[\int_{-\infty}^{t} f(t)\,dt\right] - \int_0, \tag{8.22}$$

or

$$\mathcal{L}\left[\int_{-\infty}^{t} f(t)\, dt\right] = \frac{F(s) + \int_{0}}{s}. \tag{8.23}$$

4. *Translation property*

$$\mathcal{L}[f(t - \tau)u(t - \tau)] = F(s)\epsilon^{-\tau s}, \tag{8.24}$$

where

$$F(s) = \mathcal{L}[f(t)u(t)]. \tag{8.25}$$

(See Problem 8.3.)

## 8.2 LAPLACE TRANSFORMATION OF NETWORK EQUATIONS

Let us now apply the differentiation and integration properties of Laplace transforms to the terminal characteristics of inductors and capacitors. Denoting the transforms of variables by capital letters, we have, for an inductor,

$$v_L(t) = L\frac{di_L(t)}{dt}, \tag{8.26}$$

or

$$V_L(s) = L[sI_L(s) - i_L(0)]. \tag{8.27}$$

Here $i_L(0)$ is the initial current in the inductor, evaluated at $t = 0+$. For a capacitor we may write

$$v_c(t) = \frac{1}{C}\int_{-\infty}^{t} i_c(\tau)\, d\tau, \tag{8.28}$$

or

$$V_c(s) = \frac{1}{C}\left[\frac{I_c(s)}{s} + \frac{\int_{0}}{s}\right] = \frac{I_c(s)}{Cs} + \frac{v_c(0)}{s}, \tag{8.29}$$

where $v_c(0)$ is $1/C \int_{0}$, or the initial voltage across the capacitor evaluated at $t = 0+$. Of course, we may write the terminal characteristics the other way around—i.e.,

$$i_L(t) = \frac{1}{L}\int_{-\infty}^{t} v_L(\tau)\, d\tau, \tag{8.30}$$

giving

$$I_L(s) = \frac{1}{L}\left[\frac{V_L(s)}{s} + \frac{\int_0}{s}\right] = \frac{V_L(s)}{Ls} + \frac{i_L(0)}{s}$$

$$= \frac{V_L(s)}{Ls} + \frac{\psi_L(0)}{Ls}, \tag{8.31}$$

where $\psi_L(0)$ is the initial flux linkage in the inductor, and

$$i_c(t) = C\frac{dv_c(t)}{dt}, \tag{8.32}$$

$$I_c(s) = C[sV_c(s) - v_c(0)]$$

$$= CsV_c(s) - q(0), \tag{8.33}$$

where $q(0)$ is the initial capacitor charge.

### Example 8.2

At time $t = 0$, the switch in Figure 8.1 is thrown in the direction indicated by the arrow. We will find the transform of the mesh current for $t > 0$.

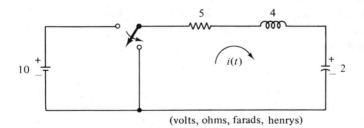

(volts, ohms, farads, henrys)

**Figure 8.1** Circuit of Example 8.2

For $t > 0$, the mesh equation is

$$4\frac{di}{dt} + 5i + \frac{1}{2}\int_{-\infty}^{t} i(\tau)\, d\tau = 0, \tag{8.34}$$

and the initial voltage across the capacitor is

$$v_c(0) \triangleq v_c(0+) = v_c(0-) = +10 \text{ V}. \tag{8.35}$$

The initial current through the inductor is

$$i(0) \triangleq i(0+) = i(0-) = 0. \tag{8.36}$$

(We have assumed the circuit to be in the constant steady state for $t < 0$.) Transforming equation 8.34, using the linearity, differentiation, and integration properties, we get

$$4[sI(s) - i(0)] + 5I(s) + \frac{I(s)}{2s} + \frac{v_c(0)}{s} = 0 \qquad (8.37)$$

or, with the initial conditions inserted,

$$\left(4s + 5 + \frac{1}{2s}\right)I(s) = -\frac{10}{s}. \qquad (8.38)$$

If we solve this transform equation algebraically for $I(s)$, we obtain

$$I(s) = \frac{-10}{4s^2 + 5s + \frac{1}{2}}. \qquad (8.39)$$

To find $i(t)$ we must find the inverse transform of $I(s)$—this will be discussed in the next section.

Example 8.2 illustrates all the characteristic features of the Laplace transform method. In particular, the original differential-integral equations are transformed into algebraic equations, which are then solved algebraically for the transform of the unknown variable. The initial conditions are accounted for at the time that the equations are transformed and need not be considered subsequently. When the response transform is found, it will be that of the complete response, including both transient and steady-state components.

### Example 8.3

The switch in the circuit of Figure 8.2(a) is closed at $t = 0$. We will use Laplace transforms to find $i(t)$ for $t > 0$.

The initial condition is

$$i(0) \triangleq i_L(0+) = i_L(0-) = 0. \qquad (8.40)$$

The effect of closing the switch is to apply a step of voltage to the $RL$ series

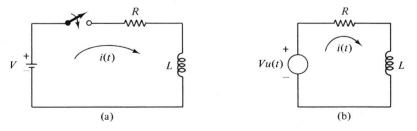

(a)  (b)

**Figure 8.2**  Circuit for Example 8.3

combination; hence the circuits indicated in parts (a) and (b) of the figure are exactly equivalent, and the differential equation for the current is

$$L\frac{di}{dt} + Ri = Vu(t). \tag{8.41}$$

The Laplace transform of this equation is

$$L[sI(s) - i(0)] + RI(s) = \frac{V}{s} \tag{8.42}$$

or

$$(Ls + R)I(s) = \frac{V}{s}. \tag{8.43}$$

Hence

$$I(s) = \frac{V}{L}\frac{1}{s[s + (R/L)]}. \tag{8.44}$$

This transform does not appear in our short table (Table 8.1). We may, however, write

$$\frac{1}{s[s + (R/L)]} = \frac{A}{s} + \frac{B}{s + (R/L)}, \tag{8.45}$$

where $A$ and $B$ are constants that may be evaluated by putting the two terms on the right side of the equation over a common denominator and then equating numerators. This gives

$$As + \frac{AR}{L} + Bs = 1, \tag{8.46}$$

from which it follows that

$$A = -B = \frac{L}{R}. \tag{8.47}$$

Hence

$$I(s) = \frac{V}{R}\left[\frac{1}{s} - \frac{1}{s + (R/L)}\right]. \tag{8.48}$$

Each of the two components in equation 8.48 is in Table 8.1, and we may therefore write the inverse transform:

$$i(t) = \frac{V}{R}[u(t) - \epsilon^{-(R/L)t}u(t)]$$

$$= \frac{V}{R}[1 - \epsilon^{-(R/L)t}]u(t). \tag{8.49}$$

The same result was obtained, with somewhat less effort, by the classical method discussed in Chapter 4, so this simple example does not really illustrate the power of the Laplace transform method. It does, however, illustrate the general procedure, which is quite different from the classical method, especially in the handling of the initial conditions.

### Example 8.4

The switch in the circuit of Figure 8.3 is thrown in the arrow direction at $t = 0$, the circuit having previously been in the constant steady state. We will find the Laplace transform of $v_2(t)$.

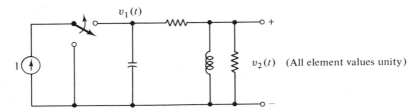

**Figure 8.3** Circuit for Example 8.4

At $t = 0-$ the charge on the capacitor and current in the inductor are both zero, and these conditions hold also at $t = 0+$. The effect of throwing the switch is to apply a unit step function of current, and the two node voltage equations may be written (using Kirchhoff's current law) as

$$\frac{dv_1}{dt} + v_1 - v_2 = u(t) \tag{8.50}$$

$$-v_1 + 2v_2 + \int_{-\infty}^{t} v_2(\tau)\, d\tau = 0. \tag{8.51}$$

If we transform these equations, with zero initial conditions, we get

$$(s + 1)V_1(s) - V_2(s) = \frac{1}{s} \tag{8.52}$$

$$-V_1(s) + \left(2 + \frac{1}{s}\right)V_2(s) = 0. \tag{8.53}$$

If we multiply both equations by $s$ and solve for $V_2(s)$ using determinants, we get

$$V(s) = \frac{\begin{vmatrix} s(s+1) & 1 \\ -s & 0 \end{vmatrix}}{\begin{vmatrix} s(s+1) & -s \\ -s & 2s+1 \end{vmatrix}} = \frac{1}{2s^2 + 2s + 1}. \tag{8.54}$$

This transform, as written, does not appear in Table 8.1. It may, however, be rewritten as

$$V(s) = \frac{1}{2s^2 + 2s + 1} = \frac{1/2j}{(s + \frac{1}{2} - j\frac{1}{2})} + \frac{-1/2j}{(s + \frac{1}{2} + j\frac{1}{2})}.$$

(That this expression is correct may be verified by putting the sum of fractions over the common denominator. In Section 8.3 we will describe a method that may be used to break up a fraction such as the preceding one into the sum of simpler fractions.) Each of the two simple fractions is now of the general form

$$F(s) = \frac{c}{s + \alpha},$$

which is one of the basic transforms in Table 8.1. Hence[1]

$$f(t) = \frac{1}{2j}\epsilon^{-[1/2+j(1/2)]t} - \frac{1}{2j}\epsilon^{[-(1/2)-j(1/2)]t}$$

$$= \epsilon^{-t/2}\frac{\epsilon^{j(t/2)} - \epsilon^{-j(t/2)}}{2j} = \epsilon^{-t/2}\sin\frac{t}{2}.$$

### Example 8.5

In the circuit of Figure 8.4, the switch is closed at $t = 0$, and there is an initial voltage of $+6$ V across the capacitor. We will find $V(s)$.

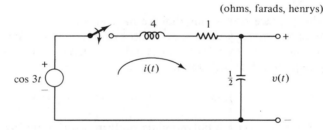

Figure 8.4   Circuit for Example 8.5

The circuit equations for $t > 0$ are

$$4\frac{di}{dt} + i + 2\int_{-\infty}^{t} i(\tau)\,d\tau = \cos 3t\, u(t) \tag{8.55}$$

$$v(t) = 2\int_{-\infty}^{t} i(\tau)d\tau. \tag{8.56}$$

The transformed equations are $[i_L(0+) = 0]$

$$4sI(s) + I(s) + \frac{2I(s)}{s} + \frac{6}{s} = \frac{s}{s^2 + 9} \tag{8.57}$$

$$V(s) = \frac{2I(s)}{s} + \frac{6}{s}. \tag{8.58}$$

---

[1]This function should really be multiplied by $u(t)$ to indicate that it is zero for negative time, as in Example 8.3. For simplicity, however, we will occasionally drop the $u(t)$, with the understanding that all $f(t)$ in this chapter are zero for negative time.

These equations may be solved as follows. From equation 8.58,

$$I(s) = 0.5sV(s) - 3, \qquad (8.59)$$

and substituting this into equation 8.57 yields

$$(2s^2 + 0.5s + 1)V(s) = \frac{s}{s^2 + 9} + 12s + 3, \qquad (8.60)$$

or

$$V(s) = \frac{12s^3 + 3s^2 + 109s + 27}{(s^2 + 9)(2s^2 + 0.5s + 1)}. \qquad (8.61)$$

As in the previous example, this transform does not directly appear in Table 8.1. We will how discuss a general method for expressing such fractions in terms of a sum of simpler fractions.

## 8.3  PARTIAL FRACTION EXPANSION OF RATIONAL FUNCTIONS

All the Laplace transforms that we have encountered in this chapter so far have been *rational functions*, i.e., ratios of finite polynomials in the variable $s$. In fact, if we transform our standard differential input–output relationship (equation 5.1), we obtain

$$M(s)Y(s) + \phi_1(s) = N(s)X(s) + \phi_2(s), \qquad (8.62)$$

where $\phi_1(s)$ and $\phi_2(s)$ are polynomials resulting from nonzero initial values of $x(t)$ and $y(t)$ and their derivatives. Solving this equation for $Y(s)$ gives

$$Y(s) = \frac{N(s)X(s) + \phi_2(s) - \phi_1(s)}{M(s)}, \qquad (8.63)$$

or

$$Y(s) = \frac{P(s)}{Q(s)}, \qquad (8.64)$$

where $P(s)$ and $Q(s)$ are just polynomials in $s$. For the time being, we will consider that the order (or degree) of the numerator polynomial $P(s)$ is at least one less than the order of the denominator $Q(s)$, which is usually the case, for reasons which we shall see shortly.

### Distinct Roots

We wish now to expand the rational function of equation 8.64 into the sum of simpler rational functions whose inverse Laplace transforms we know.

First, let us consider the case where $Q(s)$ has distinct roots (no repeated roots). In that case we may expand $Y(s)$ as

$$Y(s) = \frac{P(s)}{Q(s)} = \frac{P(s)}{(s - s_1)(s - s_2) \ldots (s - s_n)}$$

$$= \frac{a_1}{s - s_1} + \frac{a_2}{s - s_2} + \ldots + \frac{a_n}{s - s_n}, \qquad (8.65)$$

where $a_1, a_2, \ldots, a_n$ are constants (possibly complex). An expansion in this form is called a *partial fraction expansion*. If we can find the coefficients $a_1, \ldots, a_n$, then, using the second transform pair in Table 8.2, we may immediately write the inverse transform:

$$y(t) = \sum_{i=1}^{n} a_i \epsilon^{s_i t}. \qquad (8.66)$$

(Here, of course, we make use of the linearity property of Laplace transforms.) If we examine equation 8.65 we see that we may easily find any coefficient $a_i$ by simply multiplying the equation by $s - s_i$, and letting $s$ take on the value $s_i$; i.e.,

$$a_i = \frac{P(s)(s - s_i)}{Q(s)}\bigg|_{s = s_i}. \qquad (8.67)$$

[Multiplying by $(s - s_i)$ removes the denominator of the $a_i$ term, and letting $s = s_i$ causes all the other terms on the right side of equation 8.65 to be zero, so that only $a_i$ is left.]

### Example 8.6

Find $y(t)$ if $Y(s) = (s + 1)/(s^2 + 5s + 6)$.

$$Y(s) = \frac{s + 1}{(s + 2)(s + 3)} = \frac{a_1}{s + 2} + \frac{a_2}{s + 3} \qquad (8.68)$$

$$a_1 = \frac{s + 1}{s + 3}\bigg|_{s = -2} = -1 \qquad (8.69)$$

$$a_2 = \frac{s + 1}{s + 2}\bigg|_{s = -3} = +2. \qquad (8.70)$$

Therefore,

$$Y(s) = \frac{-1}{s + 2} + \frac{2}{s + 3}. \qquad (8.71)$$

or

$$y(t) = -\epsilon^{-2t} + 2\epsilon^{-3t}. \qquad (8.72)$$

*Example 8.7*

Find $y(t)$ if $Y(s) = 1/(s + 1)(s^2 + 2s + 2)$.

$$Y(s) = \frac{1}{(s + 1)(s + 1 + j1)(s + 1 - j1)}$$

$$= \frac{a_1}{s + 1} + \frac{a_2}{s + 1 + j1} + \frac{a_3}{s + 1 - j1} \tag{8.73}$$

$$a_1 = \frac{1}{(s + 1 + j1)(s + 1 - j1)}\bigg|_{s=-1} = 1 \tag{8.74}$$

$$a_2 = \frac{1}{(s + 1)(s + 1 - j1)}\bigg|_{s=-1-j1} = -\frac{1}{2} \tag{8.75}$$

$$a_3 = \frac{1}{(s + 1)(s + 1 + j1)}\bigg|_{s=-1+j1} = -\frac{1}{2}. \tag{8.76}$$

Hence

$$y(t) = \epsilon^{-t} - \tfrac{1}{2}\epsilon^{-t-jt} - \tfrac{1}{2}\epsilon^{-t+jt}$$
$$= \epsilon^{-t}[1 - \tfrac{1}{2}(\epsilon^{jt} + \epsilon^{-jt})]$$
$$= \epsilon^{-t}(1 - \cos t). \tag{8.77}$$

*Example 8.8*

Find $y(t)$ if $Y(s) = (s + 3)/(s^2 + 2s + 2)$.

$$Y(s) = \frac{s + 3}{(s + 1 + j1)(s + 1 - j1)} = \frac{a_1}{s + 1 + j1} + \frac{a_2}{s + 1 - j1} \tag{8.78}$$

$$a_1 = \frac{s + 3}{s + 1 - j1}\bigg|_{s=-1-j1} = \frac{1}{2} + j1 \tag{8.79}$$

$$a_2 = \frac{s + 3}{s + 1 + j1}\bigg|_{s=-1+j1} = \frac{1}{2} - j1. \tag{8.80}$$

Thus

$$y(t) = (\tfrac{1}{2} + j1)\epsilon^{-(1+j1)t} + (\tfrac{1}{2} - j1)\epsilon^{-(1-j1)t}$$
$$= \epsilon^{-t}[(\tfrac{1}{2} + j1)\epsilon^{-jt} + (\tfrac{1}{2} - j1)\epsilon^{jt}]. \tag{8.81}$$

We may also write $y(t)$ in sinusoidal form by using Euler's relationships, or comparing with equations 4.39 and 4.41:

$$y(t) = \sqrt{5}\,\epsilon^{-t} \cos (t - \tan^{-1} 2). \tag{8.82}$$

Note that in Example 8.8 (and, for that matter, in Example 8.7 also), the coefficients associated with conjugate complex roots are themselves conjugates of each other. This must be true, as we have seen previously, if the corresponding time functions are to be real. This fact simplifies the partial fraction expansion somewhat, since one need calculate only one of the pair of coefficients belonging to a pair of conjugate complex roots.

### Repeated Roots

If $Q(s)$ has repeated roots, we must modify the partial fraction procedure slightly. Suppose, for example, that $Q(s)$ has a double root and we label it $s_1$. Then the partial fraction expansion can take the form

$$Y(s) = \frac{P(s)}{Q(s)} = \frac{P(s)}{(s - s_1)^2(s - s_3) \ldots (s - s_n)}$$

$$= \frac{a_1}{(s - s_1)^2} + \frac{a_2}{s - s_1} + \frac{a_3}{s - s_3} + \ldots + \frac{a_n}{s - s_n}. \qquad (8.83)$$

Examination of this expression makes it evident that we may evaluate $a_1$ from

$$a_1 = \frac{P(s)(s - s_1)^2}{Q(s)}\bigg|_{s=s_1}. \qquad (8.84)$$

It is also clear, however, that $a_2$ *cannot* be obtained by the formula previously used. That is,

$$a_2 \neq \frac{P(s)(s - s_1)}{Q(s)}\bigg|_{s=s_1}. \qquad \text{(Why?)} \qquad (8.85)$$

To evaluate $a_2$ we need a different procedure. One procedure is as follows. Let us first multiply $Y(s)$ by $(s - s_1)^2$. This gives

$$\frac{P(s)(s - s_1)^2}{Q(s)}$$

$$= a_1 + a_2(s - s_1) + \frac{a_3(s - s_1)^2}{s - s_3} + \ldots + \frac{a_n(s - s_1)^2}{s - s_n}. \qquad (8.86)$$

If we next differentiate this with respect to $s$ we get

$$\frac{d}{ds}\left[\frac{P(s)(s - s_1)^2}{Q(s)}\right]$$

$$= a_2 + \ldots + \frac{2a_n(s - s_n)(s - s_1) - a_n(s - s_1)^2}{(s - s_n)^2}. \qquad (8.87)$$

Now, if we set $s = s_1$, all the terms on the right side of equation 8.87 disappear except $a_2$, hence

$$a_2 = \left[\frac{d}{ds}\frac{P(s)(s - s_1)^2}{Q(s)}\right]_{s=s_1}. \qquad (8.88)$$

All the remaining coefficients, $a_3 \ldots a_n$, may be found in the usual way, just as if a repeated root was not present.

If a root of multiplicity 3 is present, our partial fraction expansion takes the form

$$\frac{P(s)}{Q(s)} = \frac{P(s)}{(s - s_1)^3(s - s_4) \dots (s - s_n)}$$

$$= \frac{a_1}{(s - s_1)^3} + \frac{a_2}{(s - s_1)^2} + \frac{a_3}{(s - s_1)} + \frac{a_4}{s - s_4} + \dots + \frac{a_n}{s - s_n}. \quad (8.89)$$

The value of $a_1$ may be obtained from

$$a_1 = \frac{P(s)(s - s_1)^3}{Q(s)}\bigg|_{s=s_1}. \quad (8.90)$$

To find $a_2$ we must differentiate:

$$a_2 = \left[\frac{d}{ds}\frac{P(s)(s - s_1)^3}{Q(s)}\right]_{s=s_1}, \quad (8.91)$$

and to find $a_3$ we must differentiate twice:

$$a_3 = \left[\frac{1}{2}\frac{d^2}{ds^2}\frac{P(s)(s - s_1)^3}{Q(s)}\right]_{s=s_1} \quad (8.92)$$

The remainder of the coefficients may be found as usual. This procedure may be extended to roots of any multiplicity, but double-order roots are much more common in practice than roots of higher multiplicity.

### Example 8.9

Find $y(t)$ if $Y(s) = (s + 1)/(s + 2)^2(s + 3)$.

$$Y(s) = \frac{a_1}{(s + 2)^2} + \frac{a_2}{(s + 2)} + \frac{a_3}{s + 3} \quad (8.93)$$

$$a_1 = \frac{s + 1}{s + 3}\bigg|_{s=-2} = -1 \quad (8.94)$$

$$a_2 = \left[\frac{d}{ds}\frac{s + 1}{s + 3}\right]_{s=-2} = \left[\frac{(s + 3) - (s + 1)}{(s + 3)^2}\right]_{s=-2} = 2 \quad (8.95)$$

$$a_3 = \frac{s + 1}{(s + 2)^2}\bigg|_{s=-3} = -2. \quad (8.96)$$

Therefore,

$$Y(s) = \frac{-1}{(s + 2)^2} + \frac{2}{s + 2} + \frac{2}{s + 3}. \quad (8.97)$$

Using the inverse transform for multiple-order roots as given by transform 8 in Table 8.1:

$$y(t) = -t\epsilon^{-2t} + 2\epsilon^{-2t} - 2\epsilon^{-3t} = (2 - t)\epsilon^{-2t} - 2\epsilon^{-3t}. \quad (8.98)$$

Note that *single-order roots* imply *exponential terms* in the time function (possibly imaginary or complex exponentials), whereas *multiple-order roots* imply *exponentials multiplied by powers of t*. It is easy to see from the partial fraction expansion, in fact, that *these are the only time-function components that can result as a solution of our basic input–output differential equation if the order of N(s) is less than that of M(s)*.

## 8.4 WAVEFORM SYNTHESIS USING THE TRANSLATION THEOREM

If $\mathcal{L}[f(t)u(t)] = F(s)$, then, as we have already seen,

$$\mathcal{L}[f(t - \tau)u(t - \tau)] = F(s)\epsilon^{-\tau s}.$$

By using this property, and linearity, we can immediately write the Laplace transforms of numerous pulse shapes and waveforms which are of considerable engineering importance. This is illustrated by means of some examples.

### Example 8.10

Figure 8.5 shows some pulse shapes for which we desire to write the Laplace transforms. On the left is the pulse itself, on the right its constituent parts, and below each part of the figure are written the time function and transform. Additional examples of pulse synthesis are given in the problems at the end of the chapter.

### Synthesis of Semiperiodic Waveforms

A function that is zero for negative time but periodic along the positive time axis is often called a *semiperiodic function*. If such a function has a period $T$, its transform may be written

$$F(s) = \int_0^\infty f(t)\epsilon^{-st}\,dt = \int_0^T f(t)\epsilon^{-st}\,dt + \int_T^{2T} f(t)\epsilon^{-st}\,dt$$
$$+ \int_{2T}^{3T} f(t)\epsilon^{-st}\,dt + \dots. \tag{8.99}$$

If we let

$$F_0(s) \triangleq \int_0^T f(t)\epsilon^{-st}\,dt, \tag{8.100}$$

then

$$F(s) = F_0(s)(1 + \epsilon^{-sT} + \epsilon^{-2sT} + \epsilon^{-3sT} + \dots)$$
$$= F_0(s)\frac{1}{1 - \epsilon^{-sT}}. \tag{8.101}$$

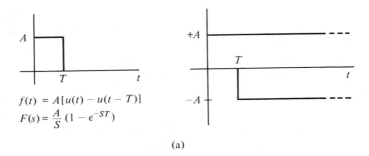

$$f(t) = A[u(t) - u(t - T)]$$

$$F(s) = \frac{A}{S}(1 - \epsilon^{-ST})$$

(a)

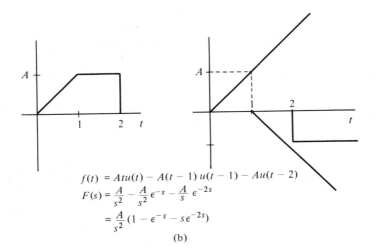

$$f(t) = Atu(t) - A(t - 1)\,u(t - 1) - Au(t - 2)$$

$$F(s) = \frac{A}{s^2} - \frac{A}{s^2}\epsilon^{-s} - \frac{A}{s}\epsilon^{-2s}$$

$$= \frac{A}{s^2}(1 - \epsilon^{-s} - s\epsilon^{-2s})$$

(b)

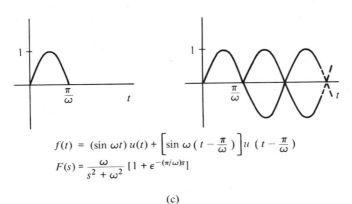

$$f(t) = (\sin \omega t)\,u(t) + \left[\sin \omega \left(t - \frac{\pi}{\omega}\right)\right]u\left(t - \frac{\pi}{\omega}\right)$$

$$F(s) = \frac{\omega}{s^2 + \omega^2}[1 + \epsilon^{-(\pi/\omega)s}]$$

(c)

**Figure 8.5** Examples of Pulse Synthesis

**264**

This result follows from the fact that our infinite series is just the binomial expansion of $(1 - \epsilon^{-sT})^{-1}$. This result allows us to write immediately a closed-form expression for the transform of any semiperiodic function if we know the transform of the function over one period.

### Example 8.11

We will find the Laplace transform of the semi-infinite rectangular wave shown in Figure 8.6. The expression for the function over its first period may be written

$$f_0(t) = u(t) - 2u(t - T) + u(t - 2T). \qquad (8.102)$$

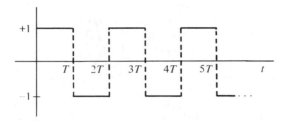

**Figure 8.6** Semi-infinite Rectangular Waveform

This has the transform

$$F_0(s) = \frac{1}{s} - \frac{2}{s}\epsilon^{-Ts} + \frac{1}{s}\epsilon^{-2Ts}. \qquad (8.103)$$

The entire waveform therefore has as its transform

$$
\begin{aligned}
F(s) &= F_0(s)\frac{1}{1 - \epsilon^{-2Ts}} \\
&= \frac{1 - 2\epsilon^{-Ts} + \epsilon^{-2Ts}}{s(1 - \epsilon^{-2Ts})} = \frac{(1 - \epsilon^{-Ts})^2}{s(1 - \epsilon^{-Ts})(1 + \epsilon^{-Ts})} \\
&= \frac{1}{s}\frac{1 - \epsilon^{-Ts}}{1 + \epsilon^{-Ts}} = \frac{1}{s}\tanh\frac{Ts}{2}.
\end{aligned}
\qquad (8.104)
$$

Additional examples will be found in the problems at the end of the chapter.

## 8.5 THE IMPULSE FUNCTION REVISITED

Let us find the Laplace transform of the unit impulse, $\delta(t)$. (Since we want our impulse to occur within the interval $(0, \infty)$—otherwise, its transform would be zero—we shall assume that it occurs at $t = 0+$.) The transform of $\delta(t)$ is

$$F(s) = \int_0^\infty \delta(t)\epsilon^{-st}\, dt = 1. \qquad (8.105)$$

Thus the Laplace and Fourier transforms of the impulse function are identical. If the impulse occurs at $t = \tau$, then

$$F(s) = \int_0^\infty \delta(t - \tau)\epsilon^{-st}\, dt = \epsilon^{-s\tau}, \qquad (8.106)$$

a result that we could also have obtained by applying the translation property.

Now let us consider the functions that are obtained by starting with the unit impulse and successively integrating:

| $f(t)$ | $F(s)$ |
|--------|--------|
| $\delta(t)$ | $1$ |
| $u(t)$ | $1/s$ |
| $tu(t)$ | $1/s^2$ |
| $\frac{1}{2}t^2u(t)$ | $1/s^3$ |
| . | . |
| . | . |
| . | . |

Successively dividing the transform of the impulse by $s$ yields the transforms of the step, ramp, parabolic function, and so on. If we go the other way, and successively multiply by $s$, we should obtain the transforms of successive derivatives of the impulse function. Of course, the impulse itself is a somewhat pathological mathematical function, and to talk about its derivatives makes even less mathematical sense than talking about the impulse being the derivative of the step function. Nevertheless, we may consider a function which, in the limit, behaves as if it were the derivative of the unit impulse. Such a function is shown in Figure 8.7; as $T \to 0$ for the function shown in part (a) of the figure, we approach in the limit a function called a *unit doublet*, which is often indicated as in part (b) of the figure. (Note that we have defined our doublet for $t = 0+$.) As in the case of the impulse function, the exact shape of the approximating function is immaterial in the limit. This "strange

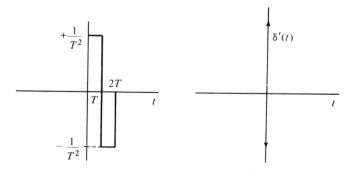

**Figure 8.7**   Unit Doublet

beast" of a function has the transform $F(s) = s$. We may, in fact, define even stranger functions by continuing to differentiate, and these functions will have transforms given by $s^2$, $s^3$, and so forth. Of these pathological limiting functions, however, only the impulse (and on rare occasions the doublet) finds practical application in system analysis.

### Partial Fraction Expansions Involving Nonnegative Powers of S

In the previous discussion of partial fraction expansions, we considered only rational fractions whose numerators were of lesser degree (or order) than the denominators. If the degree of numerator polynomial is equal to or greater than that of the denominator, then, before applying the partial fraction procedure, we must first divide the denominator into the numerator, until we obtain a remainder whose numerator degree is 1 less than that of the denominator. We may then use the partial fraction procedure on this remainder. The other terms resulting from the division will be nonnegative powers of $S$ and hence will represent impulses, doublets, etc., in the time function.

#### Example 8.12

Find $f(t)$ if $F(s) = (s^3 + 4s^2 + 6s + 5)/(s^2 + 3s + 2)$.
   Division yields

$$
\begin{array}{r}
s + 1 + \dfrac{s + 3}{s^2 + 3s + 2} \\[2pt]
s^2 + 3s + 2 \overline{)s^3 + 4s^2 + 6s + 5} \\
\underline{s^3 + 3s^2 + 2s \phantom{+ 6}} \\
s^2 + 4s + 5 \\
\underline{s^2 + 3s + 2} \\
s + 3.
\end{array}
$$

We may now expand the remainder using the partial fraction method:

$$s + 1 + \frac{s + 3}{(s + 1)(s + 2)} = s + 1 + \frac{2}{s + 1} + \frac{-1}{s + 2}. \tag{8.107}$$

If we denote a unit doublet by $\delta'(t)$, we may write

$$f(t) = \delta'(t) + \delta(t) + 2\epsilon^{-t} - \epsilon^{-2t}. \tag{8.108}$$

#### Example 8.13

We will find the impulse response of the circuit shown in Figure 8.8. If $v_1(t) = \delta(t)$, we have

$$Ri(t) + \frac{1}{C} \int_{-\infty}^{t} i(\tau)\, d\tau = \delta(t) \tag{8.109}$$

$$v_2(t) = Ri(t). \tag{8.110}$$

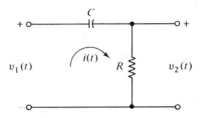

**Figure 8.8**  Circuit for Example 8.13

If we assume zero initial charge on the capacitor, the transforms of these equations are

$$RI(s) + \frac{I(s)}{Cs} = 1 \qquad (8.111)$$

$$V_2(s) = RI(s). \qquad (8.112)$$

Eliminating $I(s)$ gives

$$V_2(s) = \frac{RCs}{RCs + 1}. \qquad (8.113)$$

To find $v_2(t)$, we first divide the denominator of equation 8.113 into the numerator, to obtain

$$V_2(s) = 1 - \frac{1}{RCs + 1} = 1 - \frac{1}{RC} \frac{1}{s + (1/RC)}. \qquad (8.114)$$

Thus the inverse transform is

$$v_2(t) = \delta(t) - \frac{1}{RC}\epsilon^{-t/RC}, \qquad (8.115)$$

and is made up of an impulse function (the impulse of current charges the capacitor) and a negative decaying exponential (due to the discharge of the capacitor through the output resistor).

## 8.6  INITIAL- AND FINAL-VALUE THEOREMS

If we take the Laplace transform of the first derivative of a function $f(t)$, we obtain

$$\int_0^\infty \frac{df(t)}{dt} \epsilon^{-st} dt = sF(s) - f(0), \qquad (8.116)$$

where, as before, we will approach the lower limit from the right, so that $f(0)$ really means $f(0+)$. [This distinction is only necessary if $f(t)$ is discontinuous at the origin.] We will now take the limit of both sides of equation 8.116 as $s \longrightarrow \infty$. First, however, we should ask ourselves exactly what we mean by $s \longrightarrow \infty$, since $s$ is a complex variable. For example, we might hold the real part of $s$ constant and let the imaginary part approach infinity, or vice versa. If the real part of $s$ approaches infinity, then

$$\epsilon^{-st} = \epsilon^{-(\sigma + j\omega)t} = \epsilon^{-\sigma t}(\cos \omega t - j \sin \omega t)$$

approaches zero regardless of what the imaginary part of $s$ does. Thus, if we treat $s$ as if it were a real variable in equation 8.116 and let $s \longrightarrow \infty$, the integrand on the left side of the equation approaches zero, and we therefore have

$$f(0) = \lim_{s \to \infty} sF(s). \tag{8.117}$$

This result is called the *initial-value theorem*.

Let us now return to equation 8.116, but this time take the limit as $s \longrightarrow 0$. This gives

$$\int_0^\infty \frac{df(t)}{dt} \, dt = \lim_{s \to 0} sF(s) - f(0) = \int_0^\infty df(t) = f(\infty) - f(0), \tag{8.118}$$

where $f(\infty)$ means $\lim_{t \to \infty} f(t)$. Hence we have

$$f(\infty) = \lim_{s \to 0} sF(s), \tag{8.119}$$

and this is called the *final-value theorem*.

### Example 8.14

The Laplace transform of $\epsilon^{-\alpha t}$ is

$$F(S) = \frac{1}{s + \alpha}. \tag{8.120}$$

Let us check the initial and final values of $f(t)$, using our theorems.

$$f(0) = \lim_{s \to \infty} \frac{s}{s + \alpha} = 1 \tag{8.121}$$

$$f(\infty) = \lim_{s \to 0} \frac{s}{s + \alpha} = 0. \tag{8.122}$$

### Example 8.15

Let us check the initial and final values of $v_2(t)$ in Example 8.13. Referring to equation 8.113, we have

$$f(0) = \lim_{s \to \infty} \frac{RCs^2}{RCs + 1} = \infty \tag{8.123}$$

$$f(\infty) = \lim_{s \to 0} \frac{RCs^2}{RCs + 1} = 0. \tag{8.124}$$

Note that these results agree with the expression for $v_2(t)$ given in 8.115.

*Example 8.16*

The transform of the unit step function is $1/s$. Note that the initial and final values as given by the theorems are both unity, which is consistent with our policy of interpreting $f(0)$ as $f(0+)$.

## 8.7 LAPLACE TRANSFER FUNCTIONS

In Chapters 6 and 7 we used ratios of Fourier transforms to describe the input–output relationship of a network or system. Examples included impedances, admittances, voltage transfer ratios, current transfer ratios, and so forth. All these system functions were of the form of complex functions of the real frequency variable $\omega$. Because of the similar general nature of Laplace and Fourier transforms, it should be evident at this point that we may define similar ratios using Laplace transforms, and thereby work with system functions that are functions of the complex frequency variable $s$. When the system functions are defined in terms of Laplace transforms, the term *transfer function* is generally used to denote the ratio of output transform to input transform. Again, we will talk about impedances, admittances, voltage or current transfer ratios, etc.; these may all be thought of as particular kinds of transfer functions.

### Laplace Impedance and Admittance

Since Laplace transfer functions, like the system functions previously discussed, relate output to input transforms, and since this relationship is a unique one only if initial energy storages in the network or system are zero, *we will assume zero initial conditions whenever we use transfer function analysis.* The output of the network then is due entirely to the forcing function, and the transforms of the two are uniquely related. If we consider the terminal characteristics of inductors and capacitors, with zero initial conditions, we have the relationships shown in Table 8.2. The impedances and admittances are of the same form as the Fourier impedances and admittances, except that

**Table 8.2** $Z$ and $Y$ of Inductors and Capacitors

|  | Inductor | Capacitor |
|---|---|---|
| Differential equation | $v(t) = L\dfrac{d}{dt}i(t)$ | $v(t) = \dfrac{1}{C}\displaystyle\int_{-\infty}^{t} i(\tau)\,d\tau$ |
| Transform equation | $V(s) = LsI(s)$ | $V(s) = \dfrac{1}{Cs}I(s)$ |
| Impedance | $Z(s) = \dfrac{V(s)}{I(s)} = Ls$ | $Z(s) = \dfrac{V(s)}{I(s)} = \dfrac{1}{Cs}$ |
| Admittance | $Y(s) = \dfrac{I(s)}{V(s)} = \dfrac{1}{Ls}$ | $Y(s) = \dfrac{I(s)}{V(s)} = Cs$ |

$j\omega$ is replaced by $s$. In fact, if we compare the differentiation and integration properties of the two types of transform, it is clear that (with our assumption of zero initial conditions) any Laplace transfer function may be obtained from the corresponding Fourier system function by simply replacing ($j\omega$) by $s$, and vice versa. [It should be carefully noted that we are not making this claim for the transforms themselves but only for transfer functions, which represent ratios of transforms. If a time function is zero for negative time and absolutely integrable, then it will have both a Laplace and Fourier transform in the usual sense, and the one may be obtained from the other by replacing $s$ by $j\omega$, and vice versa. As we have seen, however, many time functions, such as growing exponentials, possess Laplace transforms but not Fourier transforms, and replacing $s$ by $j\omega$ is therefore not always valid.]

It should be clear at this point that the rules for combining Laplace impedances and admittances in series and parallel are identical to those for Fourier impedances and admittances; i.e., for series combinations,

$$Z(s) = \sum_i Z_i(s) \tag{8.125}$$

and

$$\frac{1}{Y(s)} = \sum_i \frac{1}{Y_i(s)}, \tag{8.126}$$

and for parallel combinations,

$$Y(s) = \sum_i Y_i(s) \tag{8.127}$$

and

$$\frac{1}{Z(s)} = \sum_i \frac{1}{Z_i(s)}. \tag{8.128}$$

In fact, all our previous techniques and theorems involving impedances and admittances carry over directly to the Laplace domain, including Thévenin's and Norton's theorems, superposition, the ladder method of evaluating a transfer ratio, and so forth.

### Example 8.17

Figure 8.9 shows a circuit with impedances labeled. The impedance seen looking into the input terminals (with the output terminals left open-circuited) is

$$Z_{\text{in}}(s) = Ls + R + \frac{1}{Cs} = \frac{LCs^2 + RCs + 1}{Cs}, \tag{8.129}$$

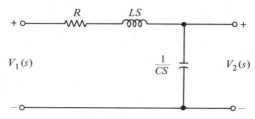

**Figure 8.9** Circuit of Example 8.17

and the voltage transfer ratio is

$$T(s) = \frac{V_2(s)}{V_1(s)} = \frac{1}{LCs^2 + RCs + 1}. \tag{8.130}$$

If the input is a unit step function, then $V_1(s) = 1/s$, and

$$V_2(s) = \frac{1}{s(LCs^2 + RCs + 1)}. \tag{8.131}$$

Let us check this, using the initial- and final-value theorems:

$$v_2(0+) = \lim_{s \to \infty} \frac{1}{LCs^2 + RCs + 1} = 0. \tag{8.132}$$

$$v_2(\infty) = \lim_{s \to 0} \frac{1}{LCs^2 + RCs + 1} = 1. \tag{8.133}$$

These results agree with our physical reasoning. It is well to apply these checks to a transfer function whenever the initial and final values of the output are known. Note that the transfer functions of equations 8.129 and 8.130 can be expressed in the Fourier domain by simply replacing $s$ by $j\omega$.

## 8.8  THE S-PLANE: POLES AND ZEROS

A specific value of the complex variable $s = \sigma + j\omega$ defines a point in the complex plane having a real coordinate $\sigma$ and an imaginary coordinate $\omega$. As we have seen, any transfer function of a lumped linear network, be it an impedance, transfer ratio, or whatever, will be a rational function of $s$; i.e., $F(s)$ will have the form of a ratio of finite polynomials in $s$. A specific value of $s$ will determine a specific value of $F(s)$—we may think in terms of a mapping from the $s$-plane to the $F(s)$-plane. In this mapping, there are certain values of $s$ which are of key importance, and in fact determine the entire character (with one minor exception) of the transfer function. These are the values of $s$ for which $F(s)$ becomes either zero or infinite. The former are the roots of the numerator of $F(s)$ and are called *zeros* of $F(s)$. The latter are the roots of the denominator of $F(s)$ and are called *poles* or $F(s)$. If we know all these roots (i.e., if we know the poles and zeros), then we know $F(s)$ to within a multiplicative constant, which is the minor exception mentioned above. In other words, we may write $F(s)$ in factored form as

$$F(s) = K\frac{(s - z_1)(s - z_2)(s - z_3) \ldots (s - z_m)}{(s - p_1)(s - p_2)(s - p_3) \ldots (s - p_n)}, \tag{8.134}$$

where $K$ is a constant, $z_1 \ldots z_m$ are the zeros of $F(s)$, and $p_1 \ldots p_n$ are the poles of $F(s)$. The usual way of indicating the poles and zeros in the $s$-plane is to use a small circle for a zero and a small x for a pole, as indicated in Example 8.18.

### Example 8.18

Consider the transfer function

$$F(s) = \frac{(s + 1)(s + 2)}{s(s^2 + 1)(s^2 + 2s + 2)}. \tag{8.135}$$

The poles and zeros of $F(s)$ are sketched in Figure 8.10.

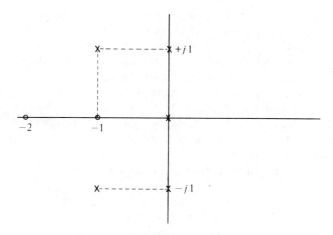

**Figure 8.10** Poles and Zeros in the $s$-plane

Let us consider what happens to the magnitude of $F(s)$ as $s$ approaches a zero or a pole value. As $s$ approaches a zero from any direction, $|F(s)| \rightarrow 0$, and as $s$ approaches a pole from any direction, $|F(s)| \rightarrow \infty$. If we visualize $|F(s)|$ plotted as a surface above the $s$-plane, with the $s$-plane itself representing $|F(s)| = 0$, then at each zero we may visualize this surface tacked down to the $s$-plane, and in the vicinity of each pole we see a mountain, as if the surface were being pushed up by a tent pole (of infinite height) at the pole position. To quote Van Valkenburg[2]: "Poles and zeros are the lifeblood of a function; without poles and zeros the function reduces to a dull, drab, grubby constant—a function which does not change under any conditions. Without poles and zeros, the three-dimensional representation of the network function becomes a tedious expanse of mathematical desert—absolutely flat. But add a few poles and a few zeros and we have a land of spectacular peaks (elevation: $\infty$) and beautiful springs (elevation: 0), the $s$ land whose coordinates (latitude and longitude with respect to $s = 0$ rather than Greenwich) are complex frequencies."

### Time Functions as Related to Pole–Zero Positions

If the input or forcing function of our lumped linear network has a transform that is a rational function (which is usually the case in practice—see Table 8.1, for example), then the output or response function is the product of two rational functions, which is another rational function. Hence the response function will also be characterized by its poles and zeros, which will be the sum total of the poles and zeros of the input function and the transfer function. Let us see how the time response is related to the poles and zeros of its transform.

In this discussion we will assume that the number of poles in the finite part of the $s$-plane exceeds the number of zeros. This will ensure that the function has a partial fraction expansion, since the numerator will have a degree at least one less than the denominator. (Actually, if we include poles and zeros at infinity, the total number of zeros will always equal the total number of poles—why?) If the number of finite zeros exceeds the number of finite poles, then we would first divide the denominator into the numerator, as previously explained, giving components that are nonnegative powers of $s$. Assuming that this has been done, if necessary, we will examine the remaining partial fraction expansion.

Our partial fraction expansion will be of the form

$$F(s) = K\frac{(s - z_1)(s - z_2) \ldots (s - z_m)}{(s - p_1)(s - p_2) \ldots (s - p_n)}$$

$$= \frac{C_1}{s - p_1} + \frac{C_2}{s - p_2} + \ldots \frac{C_n}{s - p_n}. \tag{8.136}$$

This corresponds to the time function

$$f(t) = C_1\epsilon^{+p_1 t} + C_2\epsilon^{+p_2 t} + \ldots + C_n\epsilon^{+p_n t}. \tag{8.137}$$

As we have already seen, some of the $p$'s may occur as conjugate imaginary

[2]M. E. Van Valkenburg, *Network Analysis*, 3rd ed., Prentice-Hall, Inc., Englewood Cliffs, N.J., 1974.

or complex pairs, in which case the related $C$'s will also form conjugate pairs, giving sinusoidal or damped sinusoidal time functions. Note, however, that *the basic time function character of $f(t)$ is determined by the poles of $F(s)$*; the zeros and the constant $K$ enter only into the determination of the coefficients, which determine the relative weighting of the exponential components in $f(t)$. In fact, if we consider how $F(s)$ is obtained from the differential equations of a system, it is clear that the poles of $F(s)$ are simply the roots of the characteristic equation that we discussed in Chapter 4, and the poles correspond to components of $f(t)$ as given in Table. 8.3. It is evident that if a

**Table 8.3**

| Pole Positions | $f(t)$ component ($t > 0$) |
|---|---|
| $p = -\alpha$ | $C\epsilon^{-\alpha t}$ |
| $p_1 = -\alpha$ <br> $p_2 = -\alpha$ | $C_1\epsilon^{-\alpha t} + C_2 t \epsilon^{-\alpha t}$ |
| $p_1 = +j\omega_0$ <br> $p_2 = -j\omega_0$ | $K \cos(\omega_0 t + \theta)$ |
| $p_1 = -\zeta\omega_0 + j\omega_0\sqrt{1 - \zeta^2}$ <br> $p_2 = -\zeta\omega_0 - j\omega_0\sqrt{1 - \zeta^2}$ | $K\epsilon^{-\zeta\omega_0 t} \cos[(\omega_0\sqrt{1 - \zeta^2})t + \theta]$ |

system is stable, all the poles will be in the left-hand plane—i.e., the real parts of the poles will be negative, else there will be growing (or at least non-decaying) exponentials in $f(t)$. This is discussed in greater detail in Section 8.9.

### Frequency Response as Related to Pole–Zero Positions

To obtain the Fourier system function $F(j\omega)$ from the Laplace transfer function $F(s)$, we need only to replace the variable $s$ by $j\omega$. In terms of the $s$-plane, this means that we consider the behavior of $F(s)$ *only along the $j\omega$ axis.* Suppose that we desire an amplitude frequency response, which we have defined as $|F(j\omega)|$, plotted as a function of $\omega$. If we think in terms of the surface given by $|F(s)|$ plotted over the entire $s$-plane, it is clear that our amplitude frequency response will just be the height of this surface along the $j\omega$ axis (i.e., for $s = j\omega$). The contours of the $|F(s)|$ surface are, in turn, determined by the pole–zero locations. It may help to think of this surface as a thin rubber sheet, tacked down to the $s$-plane at the zeros, and pushed up (actually to infinite height) at the poles, somewhat in the fashion of a circus tent. If a pole is very close to the $j\omega$ axis, for example, then as we go past this pole on our way up the $j\omega$ axis, $|F(j\omega)|$ will increase greatly and then decrease, as we traverse the mountainside, so to speak. This is the phenomena of resonance—the closer the pole is to the $j\omega$ axis, the sharper and higher will be the resonance peak. Visualizing the $|F(s)|$ surface in this way allows us to quickly estimate the general nature of the frequency response from the pole–zero positions.

There is another way of estimating the frequency response from $F(s)$ which is capable of graphically yielding quite accurate results. If we again denote the zeros of $F(s)$ by $z_i$ and the poles by $p_i$, we may write

$$|F(j\omega)| = \left| \frac{K(j\omega - z_1) \ldots (j\omega - z_m)}{(j\omega - p_1) \ldots (j\omega - p_n)} \right|$$

$$= K \frac{|j\omega - z_1| \ldots |j\omega - z_m|}{|j\omega - p_1| \ldots |j\omega - p_n|}. \qquad (8.138)$$

The various magnitudes involved in equation 8.138 can be interpreted as the lengths of vectors, as indicated in Figure 8.11. The way that these vector lengths determine the amplitude response is obvious from equation 8.138. Note again that if a pole is very close to the $j\omega$ axis, then as we "go past" that pole, the vector from that particular pole to the point $j\omega$ suddenly becomes very short, and since that magnitude appears in the denominator of $|F(j\omega)|$, our amplitude response will again display a large resonance peak.

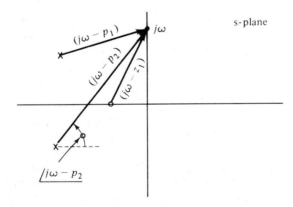

**Figure 8.11** Graphical Frequency-response Determination

If we desire the phase frequency response, we can easily obtain it by measuring the angles of the vectors from the poles and zeros to the $j\omega$ axis. From these angles we may then form

$$\underline{/F(j\omega)} = \underline{/j\omega - z_1} + \underline{/j\omega - z_2} + \ldots + \underline{/j\omega - z_m}$$

$$- \underline{/j\omega - p_1} - \underline{/j\omega - p_2} - \ldots - \underline{/j\omega - p_n}. \qquad (8.139)$$

## 8.9 STABILITY OF LINEAR SYSTEMS

The question of stability is a very important one in the analysis and design of networks and systems—particularly those which have feedback purposely built into them. It will not be our purpose here to treat this subject exhaustively or to present the several tests for stability that are covered in any good treatment of feedback control systems, as well as in many texts on network and system analysis. We will, however, provide a general definition for linear system stability and show its implications in regard to pole locations and the character of the system impulse response.

We have already noted that the poles of a lumped linear system must not be in the right half of the $s$-plane (i.e., have positive real parts), since these correspond to exponentially growing time functions. But what about poles on the $j\omega$ axis? These correspond to functions that neither grow nor decay. Also, not all linear systems are characterized by poles alone—there are other types of singularities which often occur (although not in lumped-parameter systems). In a network or system involving a time delay or transmission line, for example, a factor $\epsilon^{-sT}$ will occur in the transfer function, as we have seen (translation theorem). $\epsilon^{-sT}$ is not a rational function, and while it has a singularity (at infinity), that singularity is more complicated than a pole. Thus we need a more general definition of a stable linear system than one based entirely on pole positions.

### Linear System Stability Defined

We will say that a linear network or system is stable *if and only if its output, in response to every bounded input, is bounded.*

By $f(t)$ being bounded, we mean that there is some positive number $M$ such that

$$|f(t)| < M$$

for all $t$.

### Stability and Impulse Response

We may relate our basic definition of stability to a condition on the system impulse response. In particular, a linear system is stable *if and only if its impulse response is absolutely integrable*; i.e.,

$$\int_0^\infty |h(t)|\, dt \qquad \text{must be finite.}$$

*Proof*: If $x(t)$ and $y(t)$ are the input and output functions of the system, respectively, then by the convolution integral

$$y(t) = \int_0^\infty h(\tau)x(t - \tau)\, d\tau. \tag{8.140}$$

Replacing each factor of the integrand by its absolute value will not decrease the value of the integral; hence

$$|y(t)| \leq \int_0^\infty |h(\tau)|\,|x(t - \tau)|\, d\tau. \tag{8.141}$$

If $x(t)$ is bounded, as we assume in our basic stability definition, then

$$|y(t)| \leq \int_0^\infty |h(\tau)|M\, d\tau = M \int_0^\infty |h(\tau)|\, d\tau. \tag{8.142}$$

Hence, if $h(\tau)$ is absolutely integrable, $y(t)$ is bounded, which proves the sufficiency part of our theorem.

To show the necessity of our condition, let us suppose that $h(t)$ is not absolutely integrable. Then we may always find an input $x(t)$ that will cause $y(t)$ to be unbounded. For example, let us choose

$$x(t_1 - \tau) = \text{sgn}\,[h(\tau)] \tag{8.143}$$

for a particular $t_1$. This function is defined by

$$\text{sgn}\,[h(\tau)] = \begin{cases} +1 & \text{if } h(\tau) > 0 \\ -1 & \text{if } h(\tau) < 0 \end{cases}. \tag{8.144}$$

For this particular input,

$$y(t_1) = \int_0^\infty |h(\tau)|\, d\tau, \tag{8.145}$$

which we have assumed infinite. Hence our condition on the impulse response is both necessary and sufficient for stability.

### Stability and Pole Locations

For a *lumped-parameter* linear system, the only singularities of any transfer function are its poles, and we may state that such a system is stable *if and only if there are no poles in the right half-plane or on the $j\omega$ axis.*

*Proof*: For a transfer function $H(s)$,

$$|H(s)| = \left| \int_0^\infty h(t)\epsilon^{-st}\, dt \right| \leq \int_0^\infty |h(t)|\,|\epsilon^{-st}|\, dt. \tag{8.146}$$

For all $s$ not in the left half-plane,

$$|\epsilon^{-st}| = |\epsilon^{-\sigma t}||\epsilon^{-j\omega t}| = |\epsilon^{-\sigma t}| \le 1 \qquad (t \ge 0). \tag{8.147}$$

Hence, for all $s$ in the right half-plane or on the $j\omega$ axis,

$$|H(s)| \le \int_0^\infty |h(t)|\,dt, \tag{8.148}$$

so that if $h(t)$ is absolutely integrable, $|H(s)|$ is bounded, which means it can have no poles for those values of $s$. This proves the necessity of the condition.

To show the sufficiency of the condition, we need only consider the partial fraction expansion of $H(s)$. If all the poles have negative real parts, then, as we have seen, all the time functions contain decaying exponentials. Even if the poles are of multiple order, we obtain only time functions of the form

$$\frac{C}{n!}t^n\epsilon^{-\alpha t}.$$

Therefore, the inverse transform of $H(s)$ (i.e., the impulse response) will be absolutely integrable, since the component functions corresponding to the various terms of the partial fraction expansion will each be absolutely integrable. Thus our condition on the pole locations is both necessary and sufficient for stability, as defined in terms of the boundedness of the output.

### 8.10 LAPLACE TRANSFORM ANALYSIS OF NONELECTRICAL SYSTEMS

It should be evident from prior discussions in this chapter that all the Laplace transform definitions and techniques are applicable to other than electrical systems. In fact, they are applicable to any systems described by ordinary linear differential equations (and, to a limited extent, partial differential equations), regardless of the physical nature of the variables in the sys-

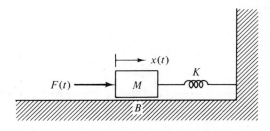

**Figure 8.12** Mass-spring System

tem. We will illustrate this by considering a lumped mechanical system consisting of a mass and spring, as indicated in Figure 8.12.

### Example 8.19

The mass $M$ is subject to a force $F(t)$, and the resulting displacement is $x(t)$. $K$ is the spring constant and $B$ the viscous damping constant, quantities that we have previously considered (equations 3.14 through 3.16). Equating the forces acting on the mass, we obtain

$$M\ddot{x} + B\dot{x} + Kx = F(t), \tag{8.149}$$

or, writing the equation in terms of the velocity $V = \dot{x}$, we have

$$M\dot{V} + BV + K \int_{-\infty}^{t} V \, d\tau = F(t). \tag{8.150}$$

If we consider $F(t)$ to be analogous to an electrical voltage, the similarity of equation 8.150 to the mesh equation

$$L\dot{i} + Ri + \frac{1}{C} \int_{-\infty}^{t} i \, d\tau = v(t) \tag{8.151}$$

is obvious. Or, if we consider $F(t)$ to be analogous to a current, we have the node equation

$$c\dot{v} + \frac{1}{R}v + \frac{1}{L} \int_{-\infty}^{t} v \, d\tau = i(t). \tag{8.152}$$

Hence either of the two circuits indicated in Figure 8.13 (which are described by equations 8.151 and 8.152, respectively) is analogous to the mechanical system. If we assume that the mechanical system starts with zero initial conditions $[x(0) = \dot{x}(0) = 0]$, then the Laplace transform of equation 8.150 is

$$MsV(s) + BV(s) + \frac{K}{s}V(s) = F(s), \tag{8.153}$$

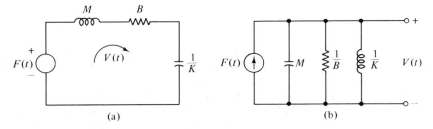

(a)  (b)

**Figure 8.13** Circuits Analogous to the Mechanical System of Figure 8.12

or

$$\frac{V(s)}{F(s)} = \frac{s}{Ms^2 + Bs + K}.$$

(8.154)

and this transfer function may be considered as either a mechanical admittance or a mechanical impedance, by analogy with the circuits of Figure 8.13(a) and (b), respectively, and the time function corresponding to this transform will be the impulse response of the mechanical system, since if $F(s) = 1$, then $V(s)$ is equal to this transfer function. If we apply the initial- and final-value theorems to this impulse response, we obtain

$$V(0) = \lim_{s \to \infty} \frac{s^2}{Ms^2 + Bs + K} = \frac{1}{M}$$

(8.155)

$$V(\infty) = \lim_{s \to 0} \frac{s^2}{Ms^2 + Bs + K} = 0,$$

(8.156)

and whether or not the impulse response is oscillatory or not will depend on the pole positions—i.e., the roots of $Ms^2 + Bs + K = 0$. In short, all the techniques discussed in this chapter apply just as well to our mechanical system as to our electrical circuits.

Any lumped, linear mechanical system containing masses, springs, and viscous dampers will have an *RLC*-circuit analog. A pair of gears, for example (or a lever), is analogous to an electrical transformer. It is not our purpose in this text to treat electrical–mechanical analogs in detail; we will, however, give one more example, similar to the previous one, but slightly more complicated.

### Example 8.20

Consider the two coupled masses of Figure 8.14(a). If we express the force balance on each of the two masses (assuming viscous damping), we obtain the following pair of equations:

$$M_1 \ddot{x}_1 + B_1 \dot{x}_1 + K_1(x_1 - x_2) = F(t)$$

(8.157)

$$M_2 \ddot{x}_2 + B_2 \dot{x}_2 + K_1(x_2 - x_1) + K_2 x_2 = 0.$$

(8.158)

If we express these equations in terms of the two velocities, we obtain

$$M_1 \dot{V}_1 + B_1 V_1 + K_1 \int_{-\infty}^{t} (V_1 - V_2)\, d\tau = F(t)$$

(8.159)

$$M_2 \dot{V}_2 + B_2 V_2 + K_1 \int_{-\infty}^{t} (V_2 - V_1)\, d\tau + K_2 \int_{-\infty}^{t} V_2\, d\tau = 0.$$

(8.160)

The analogous electrical equations, with force analogous to voltage, are

$$L_1 \frac{di_1}{dt} + R_1 i_1 + \frac{1}{c_1} \int_{-\infty}^{t} (i_1 - i_2)\, d\tau = v(t)$$

(8.161)

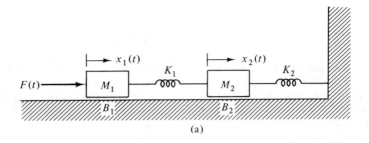

(a)

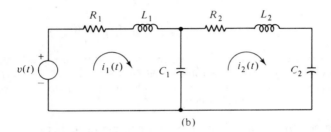

(b)

**Figure 8.14**  Lumped Mechanical System and Analogous
Electrical Circuit

$$L_2 \frac{di_2}{dt} + R_2 i_2 + \frac{1}{c_1} \int_{-\infty}^{t} (i_2 - i_1)\, d\tau + \frac{1}{c_2} \int_{-\infty}^{t} i_2\, d\tau = 0 \qquad (8.162)$$

These are seen to be the mesh equations for the two mesh circuit of Figure
8.14(b).

## PROBLEMS

**8.1** Show that the following are correct:

(a) $\mathcal{L}[u(t)] = \dfrac{1}{s}$.

(b) $\mathcal{L}[(\cos \omega_0 t)u(t)] = \dfrac{s}{s^2 + \omega_0^2}$.

(c) $\mathcal{L}[tu(t)] = \dfrac{1}{s^2}$.

**8.2** What is the abscissa of absolute convergence for the transforms of Problem
8.1?

**8.3** Show that the linearity and translation properties of Laplace transforms are
valid.

**8.4** In the network of Figure 8.15, the initial conditions at time $t = 0$ are all
zero. If $v_1(t) = t\epsilon^{-\alpha t}u(t)$, find $I_2(s)$.

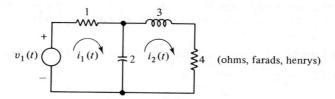

**Figure 8.15**   Circuit of Problem 8.4

**8.5**   The circuit of Figure 8.16 is in the constant steady state prior to $t = 0$. At $t = 0$ the switch is thrown as shown. Find $v(t)$ using Laplace transforms.

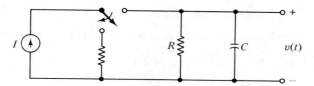

**Figure 8.16**   Circuit of Problem 8.5

**8.6**   At $t = 0$ the switch in the circuit of Figure 8.17 is thrown in the direction indicated, the circuit having been in the constant steady state prior to $t = 0$. Find $i(t)$ using Laplace transforms.

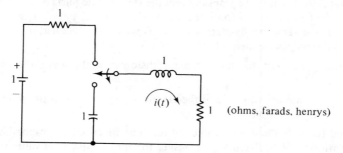

**Figure 8.17**   Circuit of Problem 8.6

**8.7**   In the circuit of Figure 8.18, $v_1(t)$ is a rectangular pulse of height $V$ volts and duration $T$ seconds. Find an expression for $v_2(t)$, using Laplace trans-

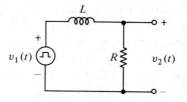

**Figure 8.18**   Circuit of Problem 8.7

forms. Sketch $v_2(t)$. Find the impulse response of the circuit by considering the limit as $V$ approaches infinity and $T$ approaches zero in such a way that $VT$ remains constant.

**8.8** Find the time functions corresponding to the following transforms:

(a) $F(s) = \dfrac{3s + 1}{s^2 + 5s + 6}$.

(b) $F(s) = \dfrac{s^3}{s^2 + 2s + 2}$.

(c) $F(s) = \dfrac{1}{s^2 + 4s + 4}$.

(d) $F(s) = \dfrac{s^3 + 2s^2 + s + 1}{s^2(s^2 + 4)}$.

(e) $F(s) = \dfrac{s + 1}{(s + 2)^2(s + 3)^2}$.

**8.9** Solve the following differential equations using Laplace transforms:
(a) $\dot{x} + 2x = (t^2 + t)u(t)$;     $x(0) = 0$.
(b) $\ddot{x} + 4x = 0$;     $x(0) = 1$, $\dot{x}(0) = 1$.
(c) $\ddot{x} + 5\dot{x} + 6x = (\sin t)u(t)$;     $x(0) = 1$, $\dot{x}(0) = 0$.

**8.10** In the circuit of Figure 8.18, let

$$v_1(t) = (\sin \omega t)u(t), \quad i_L(0) = 0.$$

Find $v_2(t)$. Identify the transient and the steady state parts of $v_2(t)$. After the transient has died out, only the steady-state part of $v_2(t)$ is left. Show that this is the same steady-state solution as would be obtained using the sinusoidal steady-state methods of Chapter 5.

**8.11** Write expressions for the Laplace transforms of the waveforms shown in Figure 8.19.

**8.12** Find the impulse response of the circuit of Figure 8.15, with $v_1(t)$ the input and $i_2(t)$ the output.

**8.13** Find the initial and final values of the time functions corresponding to the transforms of Problem 8.8, by using the initial- and final-value theorems. Check the results against those of Problem 8.8.

**8.14** Find the Laplace impedance and admittance as seen looking into each pair of terminals of the circuit of Figure 8.20, with the other pair of terminals in each case open-circuited.

**8.15** Find the resistive and reactive components of the Fourier impedance seen looking into the $v_1 - i_1$ terminal of the circuit of Figure 8.20.

**8.16** Find the Laplace voltage transfer ratio $V_2(S)/V_1(S)$ for the circuit of Figure 8.20.

**8.17** Sketch, in the $s$-plane, the pole–zero locations of the voltage transfer ratio of Problem 8.16.

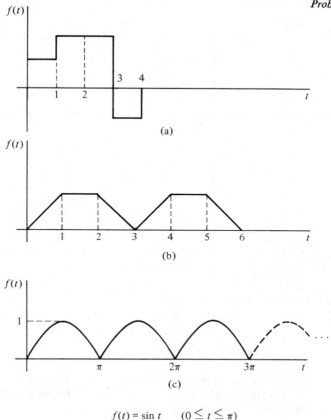

$$f(t) = \sin t \qquad (0 \le t \le \pi)$$

**Figure 8.19** Waveforms of Problem 8.11

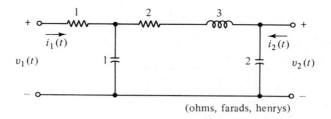

(ohms, farads, henrys)

**Figure 8.20** Circuit of Problem 8.14

**8.18** Find, to within a multiplicative constant, the time functions and Laplace transforms corresponding to the pole–zero patterns shown in Figure 8.21.

**8.19** Discuss (qualitatively) the frequency response associated with the pole–zero patterns of Figure 8.21. In particular, discuss the nature of the frequency

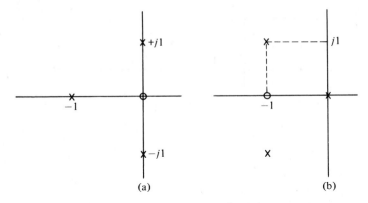

**Figure 8.21** Pole-zero Patterns for Problem 8.18

(amplitude) response at very low frequencies, very high frequencies, and wherever a pole or zero appears on or near the $j\omega$ axis.

**8.20** An amplifier has a voltage transfer ratio (gain) of $K$ (a constant). The output voltage of the amplifier is delayed by $T$ seconds and then subtracted from the input voltage. The overall network may be shown to have a transfer function

$$T(s) = \frac{K}{1 + K\epsilon^{-sT}}.$$

(Can you show this?) Find the location of the poles of $T(s)$ as a function of $K$, and discuss the stability of the network as $K$ is varied from zero to infinity. (*Hint:* What are the real and imaginary parts of $\epsilon^{-sT}$?)

# 9

# *State Variables*

In this chapter we shall examine a somewhat different way of expressing the input–output relationships of lumped linear networks and systems in the time domain, which has certain advantages over the mathematical models that we have discussed so far. The primary advantages of this new mathematical model, which we will call a *state-variable* or *state-space model*, are the following:

1. It is very convenient for analyzing circuits or systems that have multiple inputs and outputs.
2. It makes possible very simple and very standardized techniques for simulating systems on a digital computer, regardless of the order of the system or how many inputs and outputs it has.
3. It makes possible the derivation of many theoretical results (particularly in the field of optimal control) which are valid regardless of the order or complexity of the system.

In the state-variable formulation the differential or difference equations of a network or system are expressed in terms of vector and matrix quantities; an $n$th-order differential equation, for example, is written as a first-order differential equation, but the dependent variable is an $n$-dimensional vector variable, and the coefficients take the form of matrices. Hence the solution of these equations involves a number of concepts and techniques of linear algebra, and it is assumed that the reader has a rudimentary background in the algebra of vectors and matrices. For the student who desires more background on the linear algebra concepts mentioned in this chapter, a very excellent treatment may be found in Pease's book.[1]

## 9.1 STATE-VARIABLE MODEL OF A LUMPED LINEAR DIFFERENTIAL SYSTEM

Let us consider a system with multiple inputs and outputs, differentially related, as in Figure 9.1. We will define an input vector $u$ and an output vector $y$ as column vectors having the individual (scalar) inputs and outputs as components. (In this chapter we will not use a line above a variable to denote its

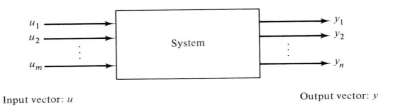

Input vector: $u$      Output vector: $y$

**Figure 9.1**   Multiple Input-output System

vector nature as we did with the spatial vectors in Chapters 1 and 2. Usually, in this chapter, a lower case letter will denote a vector, an upper case letter a matrix, and subscripts will indicate that a quantity is a scalar component of a vector or matrix.) As we shall see, we may express the relationship between the input and output vectors by the following set of two equations:

$$\dot{x}(t) = A(t)x(t) + B(t)u(t) \tag{9.1}$$

$$y(t) = C(t)x(t) + D(t)u(t). \tag{9.2}$$

$A(t)$, $B(t)$, $C(t)$, and $D(t)$ are matrices that characterize the system, and $x(t)$ is called the *state vector* of the system, for reasons that will become apparent shortly. Actually, this formulation implies that the system is a time-varying one. If the system, like most of the networks that we have considered in this text, is time-invariant, then the matrix coefficients are constant matrices and we may write

$$\dot{x}(t) = Ax(t) + Bu(t) \tag{9.3}$$

$$y(t) = Cx(t) + Du(t). \tag{9.4}$$

Let us consider now a few examples of this formulation. For the time being, we will not worry too much about how to pick the components of $x(t)$, which

[1]M. Pease, *Methods of Matrix Algebra*, Academic Press, Inc., New York, 1965.

are the state variables; a general way of doing this for any physical system will become apparent when we discuss the solution of the state equation.

**Example 9.1**

Consider the second-order differential equation

$$\ddot{v} + a\dot{v} + bv = w(t). \tag{9.5}$$

If we define

$$x_1 \triangleq v, \qquad x_2 \triangleq \dot{v},$$

then we may express equation 9.5 as

$$\dot{x}_1 = x_2 \tag{9.6}$$
$$\dot{x}_2 = -bx_1 - ax_2 + w.$$

Thus, if we define a *state vector*

$$x = \begin{bmatrix} x_1 \\ x_2 \end{bmatrix} \triangleq \begin{bmatrix} v \\ \dot{v} \end{bmatrix}$$

and a single-component *input vector* $u = w$, we have

$$\dot{x} = Ax + Bu, \tag{9.7}$$

or, writing this equation out in detail,

$$\begin{bmatrix} \dot{x}_1 \\ \dot{x}_2 \end{bmatrix} = \begin{bmatrix} 0 & 1 \\ -b & -a \end{bmatrix} \begin{bmatrix} x_1 \\ x_2 \end{bmatrix} + \begin{bmatrix} 0 \\ 1 \end{bmatrix} u. \tag{9.8}$$

So far, we have not defined an *output vector*—so we don't yet have an equation corresponding to equation 9.4. If we defined our output vector to consist of the single component $v(t)$, then we could write

$$y = \begin{bmatrix} 1 & 0 \end{bmatrix} \begin{bmatrix} x_1 \\ x_2 \end{bmatrix} + [0]u. \tag{9.9}$$

where [0] denotes the null matrix. We have now defined the $A$, $B$, $C$, and $D$ matrices associated with equation 9.5 with $v$ assumed as the output.

**Example 9.2**

Consider a series *RLC* circuit driven by a voltage source $v(t)$. The equation for the current is

$$L\frac{di}{dt} + Ri + \frac{1}{C} \int_{-\infty}^{t} i(\tau)\, d\tau = v(t). \tag{9.10}$$

Let us define the state variables as follows:

$$x_1 \triangleq i, \qquad x_2 \triangleq \frac{1}{C} \int_{-\infty}^{t} i(\tau) \, d\tau.$$

Then we have

$$\dot{x}_1 = -\frac{R}{L} x_1 - \frac{1}{L} x_2 + \frac{v}{L}$$

$$\dot{x}_2 = \frac{1}{C} x_1.$$

(9.11)

Thus again our equation is of the form

$$\dot{x} = Ax + Bu,$$

(9.12)

where

$$A \triangleq \begin{bmatrix} -\dfrac{R}{L} & -\dfrac{1}{L} \\ \dfrac{1}{C} & 0 \end{bmatrix}, \qquad B \triangleq \begin{bmatrix} \dfrac{1}{L} \\ 0 \end{bmatrix}.$$

Note that in this example we chose the inductor current and the capacitor voltage to be our state variables. The rational for this will become apparent shortly.

### Example 9.3

Consider the electrical filter network of Figure 9.2. Let us assume that $i_1(t)$ is our independent input (perhaps supplied from a previous transistor amplifier stage), and that we wish to display $v_1$, $i_1$, $v_2$, and $i_2$ as output quantities.

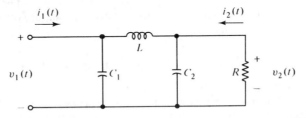

**Figure 9.2** Circuit of Example 9.3

By applying Kirchhoff's current law at the nodes on either side of the inductor, we may write the following set of equations:

$$i_1 - C_1 \dot{v}_1 - \frac{1}{L} \int_{-\infty}^{t} (v_1 - v_2) \, d\tau = 0$$

(9.13)

$$i_2 - C_2\dot{v}_2 + \frac{1}{L}\int_{-\infty}^{t} (v_1 - v_2)\, d\tau = 0 \tag{9.14}$$

$$i_2 = -\frac{v_2}{R}. \tag{9.15}$$

Following the procedure of the preceding example, we will again choose as state variables the inductor current and capacitor voltages:

$$x_1 \triangleq i_L = \frac{1}{L}\int_{-\infty}^{t} (v_1 - v_2)\, d\tau$$

$$x_2 \triangleq v_1, \qquad x_3 \triangleq v_2.$$

Hence

$$\dot{x}_1 = \frac{1}{L}(v_1 - v_2) = \frac{1}{L}x_2 - \frac{1}{L}x_3 \tag{9.16}$$

$$\dot{x}_2 = \dot{v}_1 = -\frac{1}{C_1}x_1 + \frac{1}{C_1}i_1 \tag{9.17}$$

$$\dot{x}_3 = \dot{v}_2 = \frac{1}{C_2}x_1 - \frac{1}{C_2 R}x_3. \tag{9.18}$$

Thus our equations may be expressed in the standard matrix form

$$\dot{x} = Ax + Bu \tag{9.19}$$
$$y = Cx + Du$$

by defining the matrices appropriately; i.e.,

$$
\begin{bmatrix} \dot{x}_1 \\ \dot{x}_2 \\ \dot{x}_3 \end{bmatrix}
=
\begin{bmatrix}
0 & \dfrac{1}{L} & -\dfrac{1}{L} \\
-\dfrac{1}{C_1} & 0 & 0 \\
\dfrac{1}{C_2} & 0 & -\dfrac{1}{C_2 R}
\end{bmatrix}
\begin{bmatrix} x_1 \\ x_2 \\ x_3 \end{bmatrix}
+
\begin{bmatrix} 0 \\ \dfrac{1}{C_1} \\ 0 \end{bmatrix}
[i_1]
$$

$$
\begin{bmatrix} v_1 \\ i_1 \\ v_2 \\ i_2 \end{bmatrix}
=
\begin{bmatrix}
0 & 1 & 0 \\
0 & 0 & 0 \\
0 & 0 & 1 \\
0 & 0 & -1/R
\end{bmatrix}
\begin{bmatrix} x_1 \\ x_2 \\ x_3 \end{bmatrix}
+
\begin{bmatrix} 0 \\ 1 \\ 0 \\ 0 \end{bmatrix}
[i_1]. \tag{9.20}
$$

## Example 9.4

For the sake of variety, let us consider a nonelectrical example. We will assume a point mass (of mass $M$) in free space, and for simplicity we will assume the space to be two-dimensional, described by the coordinates $(\alpha, \beta)$, as indicated in Figure 9.3. Suppose that our mass is acted upon by a force

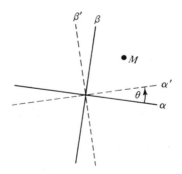

**Figure 9.3** Coordinates for Example 9.4

vector whose components are given in the $(\alpha, \beta)$ coordinate system, but for some reason we can observe only displacements in a rotated coordinate system, $(\alpha', \beta')$. These displacements in the primed coordinate system will be our outputs.

The equations of motion of the mass are

$$\ddot{\alpha} = \frac{1}{M} F_\alpha, \qquad \ddot{\beta} = \frac{1}{M} F_\beta, \tag{9.21}$$

where $F_\alpha$ and $F_\beta$ are the force vector components. If we define the state variables as

$$x_1 \triangleq \dot{\alpha}, \qquad x_2 \triangleq \dot{\beta},$$

and the output vector as

$$y \triangleq \begin{bmatrix} \dot{\alpha}' \\ \dot{\beta}' \end{bmatrix},$$

then we have the equations

$$\begin{bmatrix} \dot{x}_1 \\ \dot{x}_2 \end{bmatrix} = \begin{bmatrix} 0 \end{bmatrix} \begin{bmatrix} x_1 \\ x_2 \end{bmatrix} + \begin{bmatrix} \dfrac{1}{M} & 0 \\ 0 & \dfrac{1}{M} \end{bmatrix} \begin{bmatrix} F_\alpha \\ F_\beta \end{bmatrix}$$

$$\begin{bmatrix} y_1 \\ y_2 \end{bmatrix} = \begin{bmatrix} \cos\theta & \sin\theta \\ -\sin\theta & \cos\theta \end{bmatrix} \begin{bmatrix} x_1 \\ x_2 \end{bmatrix} + \begin{bmatrix} 0 \end{bmatrix} \begin{bmatrix} F_\alpha \\ F_\beta \end{bmatrix}. \tag{9.22}$$

These equations are now in the standard form of equations 9.3 and 9.4.

## 9.2 SOLUTIONS OF THE STATE DIFFERENTIAL EQUATION WITH TIME-VARYING COEFFICIENTS

In the preceding examples we have started with sets of $n$th-order differential equations and, by an appropriate choice of a state vector, have reduced our sets of equations to single first-order vector differential equations. If we can write a solution of a first-order vector differential equation, then we will have a solution that is independent of the order of the system—a rather intriguing possibility.

### Solution of the Homogeneous Equation

Let us consider first the more general case of a time-varying equation, and then we will return to the time-invariant case. We wish to find a solution to the equation

$$\dot{x}(t) = A(t)x(t), \tag{9.23}$$

or,

$$\dot{x}(t) - A(t)x(t) = 0. \tag{9.24}$$

It may be shown[2] that if each element of $A(t)$ is a continuous time function, then under a very general condition (called the Lipschitz condition), there will be exactly $n$ linearly independent solutions of this equation, where $n$ is the order of the $x$ vector. Thus the totality of all solutions—and there are an infinite number of them since we may choose any initial "state"—form an $n$-dimensional vector space. To specify a particular solution, one may specify a particular initial state, $x(t_0)$, and $n$ linearly independent initial state vectors will yield $n$ linearly independent solutions. Let us look at one particular set of such solutions, namely, the set $\{x_i(t)\}$ corresponding to the initial conditions:

$$x_1(t_0) = \begin{bmatrix} 1 \\ 0 \\ 0 \\ 0 \\ \cdot \\ \cdot \\ \cdot \\ 0 \end{bmatrix}, \quad x_2(t_0) = \begin{bmatrix} 0 \\ 1 \\ 0 \\ 0 \\ \cdot \\ \cdot \\ \cdot \\ 0 \end{bmatrix}, \quad x_n(t_0) = \begin{bmatrix} 0 \\ 0 \\ 0 \\ \cdot \\ \cdot \\ \cdot \\ 1 \end{bmatrix}. \tag{9.25}$$

Since linear independence of solutions implies linear independence of the set $x_i(t)$ for all values $t$, and since we have chosen $n$ vectors which are obviously independent for $t = t_0$, this set is complete and forms a basis set for our solution space. Hence any solution may be expressed as a linear combination of the elements of this set. Let us now arrange this solution set to form a square matrix, having $x_i(t)$ as the $i$th column; i.e.,

$$\phi(t, t_0) \triangleq [x_1(t)x_2(t) \ldots x_n(t)]. \tag{9.26}$$

[2]E. A. Coddington and N. Levinson, *Theory of Ordinary Differential Equations,* McGraw-Hill Book Company, New York, 1955.

From equation 2.2, at $t = t_0$, $\phi$ becomes the identity matrix,

$$\phi(t_0, t_0) = I, \tag{9.27}$$

and since each column of $\phi$ satisfies equation 9.24, so does the matrix itself (the reader may wish to verify this):

$$\frac{d}{dt} \phi(t, t_0) - A(t)\phi(t, t_0) = 0, \qquad \phi(t_0, t_0) = I. \tag{9.28}$$

It should be kept in mind that the columns of $\phi(t, t_0)$ are a very particular set of solutions to equation 9.24, namely, the solutions that result from establishing an initial condition of unity at $t = t_0$ on each of the state variables, in turn. Postmultiplying both sides of equation 2.5 by $x(t_0) = x_0$, and comparing with equation 9.24, it is evident that

$$x(t) = \phi(t, t_0)x_0, \qquad x(t_0) = x_0. \tag{9.29}$$

From equation 9.29, it is seen that $\phi(t, t_0)$, which is known as the *state transition matrix*, is the linear operator that relates the state vector $x(t)$ at time $t$ to the initial state $x_0$, and equation 9.29 may be thought of as describing the trajectory of the state vector in the state or solution space. The state transition matrix has some important properties. It is nonsingular, since by definition it is composed of linearly independent columns; hence its inverse exists for all time values. By successively applying equation 9.29, it is easily seen that

$$\phi(t_2, t_0) = \phi(t_2, t_1)\phi(t_1, t_0), \tag{9.30}$$

and by letting $t_2 = t_0$,

$$\phi(t_0, t_0) = I = \phi(t_0, t_1)\phi(t_1, t_0), \tag{9.31}$$

so

$$\phi(t_0, t_1) = \phi^{-1}(t_1, t_0). \tag{9.32}$$

Thus, if $\phi(t_1, t_0)$ transfers the system from state $x(t_0)$ to state $x(t_1)$, then $\phi^{-1}(t_1, t_0)$ operating on $x(t_1)$ brings it back again to the original state $x(t_0)$. This state transition matrix plays a fundamental role in the solution of all our system equations, as will be seen.

### Nonhomogeneous Equations

We consider now the response of the linear system to a forcing function, $u(t)$. This may most easily be done by first considering the transition matrix of another equation, which is related to equation 9.24 but is slightly different, namely,

$$\dot{z}(t) + A'(t)z(t) = 0, \tag{9.33}$$

where $A^\dagger$ is the hermitian adjoint (complex-conjugate transpose) of $A$; i.e.,

$$A^\dagger = A^{*T}, \tag{9.34}$$

where $*$ denotes complex conjugate and $T$ denotes the transpose. The equation (which is often referred to as the *adjoint equation* associated with equation 9.24) will have a transition matrix $\psi(t, t_0)$ which satisfies

$$\dot{\psi}(t, t_0) + A^\dagger(t)\psi(t, t_0) = 0, \qquad \psi(t_0, t_0) = I. \tag{9.35}$$

Let us consider the product $\psi^\dagger \phi$, and differentiate this matrix product with respect to time:

$$\frac{d}{dt}(\psi^\dagger \phi) = \dot{\psi}^\dagger \phi + \psi^\dagger \dot{\phi}. \tag{9.36}$$

From equation 9.35,

$$\dot{\psi}^\dagger = -(A^\dagger \psi)^\dagger = -\psi^\dagger A, \tag{9.37}$$

and from equation 9.28, $\dot{\phi} = +A\phi$. Hence

$$\frac{d}{dt}(\psi^\dagger \phi) = -\psi^\dagger A + \psi^\dagger A\phi \equiv 0, \tag{9.38}$$

which means that $\psi^\dagger \phi$ is a constant. Therefore,

$$\psi^\dagger(t, t_0)\phi(t, t_0) = \psi^\dagger(t_0, t_0)\phi(t_0, t_0) = I, \tag{9.39}$$

and hence

$$\psi^\dagger(t, t_0) = \phi^{-1}(t, t_0) = \phi(t_0, t). \tag{9.40}$$

We now return to the problem of solving the time-varying nonhomogeneous equation

$$\dot{x}(t) - A(t)x(t) = B(t)u(t). \tag{9.41}$$

Premultiplying both sides of this equation by $\psi^\dagger(t, t_0)$ and using equation 9.37, we have

$$\psi^\dagger(t, t_0)\dot{x}(t) - \psi^\dagger(t, t_0)A(t)x(t) = \frac{d}{dt}[\psi^\dagger(t, t_0)x(t)]$$
$$= \psi^\dagger(t, t_0)B(t)u(t)$$
$$= \phi^{-1}(t, t_0)B(t)u(t). \tag{9.42}$$

Both sides of this equation may now be integrated to yield

$$\psi^\dagger(t, t_0)x(t) = [\psi^\dagger(t, t_0)x(t)]_{t=t_0} + \int_{t_0}^{t} \phi^{-1}(\tau, t_0)B(\tau)u(\tau)\, d\tau$$
$$= x(t_0) + \int_{t_0}^{t} \phi^{-1}(\tau, t_0)B(\tau)u(\tau)\, d\tau. \tag{9.43}$$

Therefore,

$$x(t) = \phi(t, t_0)x_0 + \int_{t_0}^{t} \phi(\tau, t_0)\phi^{-1}(\tau, t_0)B(\tau)u(\tau)\,d\tau$$

$$= \phi(t, t_0)x_0 + \int_{t_0}^{t} \phi(t, \tau)B(\tau)u(\tau)\,d\tau. \tag{9.44}$$

The integral part of equation 9.44, which gives the portion of the response due to the forcing function, is a vector form of the *convolution integral* and may be thought of as a matrix of integral operators, which operates on the vector $u(t)$.

We have now formally written the solutions of both the homogeneous and "driven" time-varying differential systems in terms of the state transition matrix. [To obtain the overall output vector $y(t)$ from $x(t)$ and $u(t)$ we need only employ the simple algebraic relationship of equation 9.2—the hard work is accomplished when $x(t)$ has been found.] Of course, the problem is not entirely completed as yet; we have not yet discussed how to find $\phi$, although we have discussed its nature. To reiterate, the columns of $\phi$ are the state vector response for initial conditions of unity on each of the state variables in turn, the others having zero initial condition. Finding this particular set of solutions analytically for the time-varying-system matrix is usually difficult. If one is using a digital computer to solve the equations or simulate the system, however, there is inherently no more difficulty with the time-varying equations than with the time-invariant ones; one may simply set the initial condition on each state variable in turn to unity, and let the computer run, then generate and store each state vector solution, which will then form the columns of $\phi$. There is, however, a difference in the amount of computation and storage required between the time-varying and time-invariant cases. In the time-varying case, the columns of $\phi(t, \tau)$ must be computed not only for each $t$ in an interval of interest, but for each $\tau$ as well, while in the time-invariant case, these solutions will be invariant to the point $\tau$ at which the initial conditions are applied, as is discussed in the following section.

### 9.3 SOLUTION OF THE TIME-INVARIANT STATE DIFFERENTIAL EQUATIONS

#### Homogeneous Equation

We consider now the equation

$$\dot{x}(t) - Ax(t) = 0, \qquad x(t_0) = x_0, \tag{9.45}$$

where $A$ is now a constant matrix. Since this is simply a special case of equation 9.24, our solution again can be written in terms of the state transition matrix:

$$x(t) = \phi(t, t_0)x_0. \tag{9.46}$$

In this case we may employ any of the analytical methods applicable to time-invariant differential equations to find $\phi$; for example, if $t_0 = 0$ (which we may always assume here since our solution is invariant to translations along the time axis) and we desire the solution for $t > 0$, we may take the Laplace transform of both sides of equation 9.45, as follows:

$$sX(s) - x_0 - Ax(s) = 0 \tag{9.47}$$

$$(sI - A)X(s) = x_0 \tag{9.48}$$

$$X(s) = (sI - A)^{-1}x_0. \tag{9.49}$$

Hence $\phi(t, 0)$ is given by the inverse Laplace transform of $(sI - A)^{-1}$. An example will be presented shortly.

### Matrix Exponential

We will now give another very useful interpretation of $\phi$ when $A$ is time invariant. If equation 9.45 is a scalar equation, i.e., if there is only one component of the state vector and $A = a$, then the solution is

$$x(t) = \epsilon^{a(t-t_0)}x_0. \tag{9.50}$$

By analogy with this scalar case, one is led to defining a matrix exponential function as one that has a power-series expansion analogous to that of $\epsilon^{at}$, namely,

$$\epsilon^{At} \triangleq I + At + \ldots + \frac{A^n}{n!}t^n + \ldots. \tag{9.51}$$

This series may be differentiated term by term to yield

$$\frac{d}{dt}\epsilon^{At} = A + \dot{A}^2t + \ldots + \frac{A^{n+1}}{n!}t^n + \ldots$$

$$= A\epsilon^{At}. \tag{9.52}$$

Since $\phi(t, t_0)$ is the unique solution of

$$\dot{\phi}(t, t_0) - A\phi(t, t_0) = 0, \qquad \phi(t_0, t_0) = I, \tag{9.53}$$

it is clear that

$$\phi(t, t_0) = \epsilon^{A(t-t_0)}, \tag{9.54}$$

as may be seen by substituting equation 9.54 into 9.53 and noting that

$$\epsilon^{A(t_0-t_0)} = \epsilon^0 = I. \tag{9.55}$$

The following properties of $\epsilon^{AT}$ may be easily verified, using the series definition (9.51):

1. $\epsilon^{A(t_1 + t_2)} = \epsilon^{At_1} \epsilon^{At_2}$. $\qquad\qquad$ (9.56)
2. If $A$ and $B$ are two matrices that commute, i.e., if $AB = BA$, then

$$\epsilon^{(A+B)t} = \epsilon^{At} \epsilon^{Bt}. \qquad (9.57)$$

(This does not hold if $A$ and $B$ do not commute!)
3. $A$ and $\epsilon^{At}$ commute: i.e., $A\epsilon^{At} = \epsilon^{At}A$. (In general, however, another matrix, $B$, will not commute with $\epsilon^{At}$, unless $A = B$.)
4. $\epsilon^{At}\epsilon^{-At} = \epsilon^0 = I$; hence $(\epsilon^{At})^{-1} = \epsilon^{-At}$. $\qquad$ (9.58)
5. Since $\phi(t, 0) = \epsilon^{At}$, it is clear from equation 9.49 that the Laplace transform of $\epsilon^{At}$ is given by $(sI - A)^{-1}$, and the transform of $\epsilon^{-At}$ is $(sI + A)^{-1}$. These may be easily remembered by noting the similarity to the transform of the analogous scalar exponentials.

***Example 9.5***

$$A = \begin{bmatrix} 0 & +1 \\ -2 & -3 \end{bmatrix}.$$

Find $\epsilon^{At}$.

$$sI - A = \begin{bmatrix} s & -1 \\ 2 & s+3 \end{bmatrix} \qquad (9.59)$$

$$(sI - A)^{-1} = \begin{bmatrix} \dfrac{s+3}{(s+2)(s+1)} & \dfrac{1}{(s+2)(s+1)} \\ \dfrac{-2}{(s+2)(s+1)} & \dfrac{s}{(s+2)(s+1)} \end{bmatrix} \qquad (9.60)$$

Therefore,

$$\epsilon^{At} = \begin{bmatrix} -\epsilon^{-2t} + 2\epsilon^{-t} & -\epsilon^{-2t} + \epsilon^{-t} \\ 2\epsilon^{-2t} - 2\epsilon^{-t} & 2\epsilon^{-2t} - \epsilon^{-t} \end{bmatrix}. \qquad (9.61)$$

Hence, if this particular $A$ is our system matrix, the solution $x(t)$ is given by multiplying the initial condition vector $x(0)$ by this $\epsilon^{At}$ matrix, and if the initial conditions are at $t = t_0$, then

$$x(t) = \epsilon^{A(t-t_0)} x(t_0). \qquad (9.62)$$

***Nonhomogeneous Equations***

The solution of the nonhomogeneous time-invariant differential equation is given by equation 9.44, but now the kernal of the integral becomes simpler, since

$$\phi(t, t_0)\phi^{-1}(\tau, t_0) = \epsilon^{A(t-t_0)} \epsilon^{-A(\tau - t_0)}$$

$$= \epsilon^{A(t-\tau)} = \phi(t - \tau). \qquad (9.63)$$

Hence we may write as our general solution

$$x(t) = \epsilon^{A(t-t_0)}x(t_0) + \int_{t_0}^{t} \epsilon^{A(t-\tau)}Bu(\tau)\,d\tau. \tag{9.64}$$

## 9.4 SIGNIFICANCE OF THE STATE SOLUTIONS

Let us rewrite the time-invariant (constant coefficient) state equation solutions, assuming that $t_0 = 0$. The homogeneous equation has the solution

$$x(t) = \epsilon^{At}x(0), \tag{9.65}$$

and the nonhomogeneous equation has the solution

$$x(t) = \epsilon^{At}x(0) + \int_{0}^{t} \epsilon^{A(t-\tau)}Bu(\tau)\,d\tau. \tag{9.66}$$

It is easy to see now why $x$ is termed the state vector. $x(0)$, for example, describes the state of the system at $t = 0$, in the sense that all future states, $x(t)$, depend only on $x(0)$ and $u(\tau)$ for $0 \leq \tau \leq t$; no values of $x$ for $t < 0$ need to be considered. More generally, $x(t_1)$ describes the system state at $t = t_1$, since all future states $x(t)$ are functions of $x(t_1)$ and $u(\tau)$ over the interval $(t_1, t)$; i.e.,

$$x(t) = F[x(t_1), u(\tau)] \qquad (t_1 \leq \tau \leq t). \tag{9.67}$$

The term $x(t)$ describes the trajectory of the state vector in the $n$-dimensional state space, and $x(t_1)$ in a sense summarizes the entire past history of this trajectory, for $t < t_1$—in other words, if the system is in state $x(t_1)$ at $t = t_1$, it does not matter how it got there, insofar as its subsequent behavior is concerned. Thus "state vector" and "state variables" are logical terms to denote $x(t)$ and its components, and agree with the usual connotation of "state."

### *Choice of State Variables*

The state vector of a system is not unique—in fact, we may choose from an infinite number of different sets of state variables to describe any given network or system. For example, if $x$ is a valid state vector, and $x'$ is related to $x$ by

$$x' = Mx, \qquad x = M^{-1}x', \tag{9.68}$$

then $x'$ is also a valid state vector (Problem 9.6). Of course, any choice of variables $\{x_i\}$ such that the system equations take the form of equations 9.3 and 9.4 will be a valid choice of state variables. For a physical circuit or system, however, there is a sure-fire way of making this choice. We know that the state of a system, as defined in the preceding paragraph, will be specified if and only if all the energy storages in the system are specified. Hence any set of variables will be state variables if and only if, at any instant of time, all energy storages may be calculated from them. That is why, in previous electrical examples, we used inductor currents and capacitor voltages as the state variables—the electrical energy storages are directly related to these variables. We will illustrate this with an example involving an electro-mechanical coupled circuit.

### Example 9.6

Consider the capacitive transducer of Figure 9.4. A constant voltage ($V_0$) is applied to the electrical circuit, and a variable force is applied to the mass

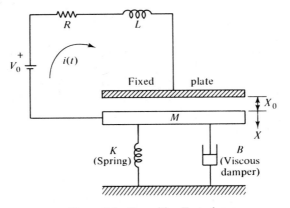

**Figure 9.4**  Capacitive Transducer

(movable capacitor plate). We will consider displacements from an equilibrium point defined by $X_0, q_0$, and $F_0$, so the total displacement, charge, and electrical force on the mass are given by $X_0 + X$, $q_0 + q$, and $F_0 + F$, respectively.

Applying Kirchhoff's voltage law to the electrical circuit yields

$$V_0 - R\frac{dq}{dt} - L\frac{d^2q}{dt^2} - \frac{1}{C(X)}(q_0 + q) = 0, \qquad (9.69)$$

and the mechanical equation of motion is

$$(F_0 + F) - M\frac{d^2X}{dt^2} - B\frac{dX}{dt} - K(X_0 + X) = 0. \qquad (9.70)$$

In addition, we have the following relations:

$$C(X) = \frac{\epsilon A}{X_0 + X} \qquad (9.71)$$

$$F_0 + F = \frac{1}{2} v_c^2 \frac{dC(X)}{dX} = -\frac{1}{2} \frac{(q + q_0)^2}{\epsilon A} \tag{9.72}$$

where $\epsilon$ is the dielectric constant, $A$ the area of the plates, and $v_c$ the capacitor voltage. We may define a set of state variables by choosing the variables directly associated with energy storages in the system; i.e.,

$$\begin{aligned}
x_1 &\triangleq q \\
x_2 &\triangleq \dot{q} \\
x_3 &\triangleq X \\
x_4 &\triangleq \dot{X}.
\end{aligned} \tag{9.73}$$

In terms of these variables our equations are

$$\begin{aligned}
\dot{x}_1 &= x_2 \\
\dot{x}_2 &= \frac{V_0}{L} - \frac{R}{L} x_2 - \frac{1}{\epsilon AL}(X_0 q_0 + X_0 x_1 + q_0 x_3 + x_1 x_3) \\
\dot{x}_3 &= x_4 \\
\dot{x}_4 &= -\frac{1}{2\epsilon AM}(x_1^2 + 2q_0 x_1 + q_0^2) - \frac{B}{M} x_4 - \frac{K}{M}(X_0 + x_3).
\end{aligned} \tag{9.74}$$

These state equations are nonlinear, because of the nonlinearity of the electro-mechanical coupling terms (equations 9.71 and 9.72). However, if the excursions of the state vector from the equilibrium point are sufficiently small, we may neglect the second-order terms $(x_1 x_3)$ and $x_1^2$, with negligible error. The equations are then in our standard linear form:

$$\dot{x} - Ax = Bu, \tag{9.75}$$

with $A$ being the matrix

$$A \triangleq \begin{bmatrix} 0 & +1 & 0 & 0 \\ \dfrac{-X_0}{\epsilon AL} & -\dfrac{R}{L} & -\dfrac{q_0}{\epsilon AL} & 0 \\ 0 & 0 & 0 & +1 \\ \dfrac{-q_0}{\epsilon AM} & 0 & -\dfrac{K}{M} & -\dfrac{B}{M} \end{bmatrix}. \tag{9.76}$$

## 9.5  ANALOG COMPUTER SIMULATION OF THE STATE MODEL

One of the big advantages of the state-variable formulation is that it describes a system in a way that allows a very simple and standardized simulation using analog computer building blocks. The linear building blocks of analog computers consist essentially of circuits capable of performing the operations of integration, addition, and multiplication by a constant. We may draw a block diagram showing the operations involved in equations 9.3 and 9.4, as indicated in Figure 9.5. In this diagram the double arrows indicate the vector variables. The box containing the integral sign represents a set of integrators—i.e., one integrator for each component of the vector

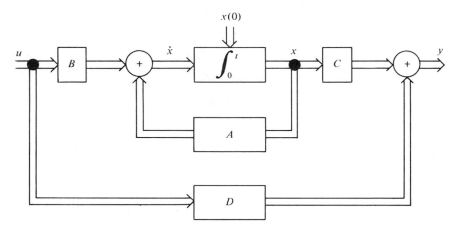

**Figure 9.5** State-variable System Model

variable $x$. The boxes containing capital letters represent matrix multiplications, which consist simply of sets of multiplications by constants, and additions. The thing that is interesting about this diagram is that it represents a simulation of any lumped, linear, time-invariant, multiple input–output system or network, regardless of the order or complexity of that system or network. Of course, an actual analog computer wiring diagram would show the details of the matrix multiplications, but, with the schematic diagram of Figure 9.5 to follow, the detailed patching of the computer is simple. Of course, the complexity and order of the system are reflected in the dimensionality of the vectors and matrices—but the basic structure indicated by Figure 9.5 is unchanged.

## 9.6 DIGITAL COMPUTER SIMULATION OF THE STATE MODEL: STATE DIFFERENCE EQUATIONS

Let us suppose that we have a linear differential network or system described by equations 9.3 and 9.4 and that we wish to simulate this system using a digital computer. There are many ways of doing this, but of course they all involve some sort of approximation. We will assume here that the input vector ($u$) is sampled at regular intervals, and we will find the state vector ($x$) and the output vector ($y$) at the sampling epochs by evaluating equation 9.64. This is analogous to the digital evaluation of the convolution integral discussed in Chapter 7.

### Piecewise Constant Input Approximation

In Chapter 7 we discussed a numerical evaluation of the convolution integral by a procedure that involved approximating our input signal between sampling epochs by straight-line segments terminating on the sample values (a trapezoidal approximation). We may do the same thing here; however, for simplicity, we will first follow an even simpler (but not quite as accurate) procedure and assume a piecewise constant (stair-step) approximation of the input vector $u$. Figure 9.6(a) shows a piecewise constant-input component, and (b) indicates one of the continuous outputs, sampled at intervals of $nT$ seconds. The input may be considered to result from a sampling operation on a continuous input function, each sample value being "held" over the interval $T$, as by a *sample-and-hold* circuit such as is often employed in sam-

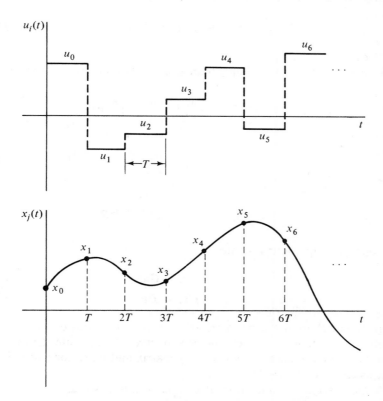

**Figure 9.6** Piecewise Constant-input Approximation and Response

pled data control systems. The state-vector components (or overall outputs) are sampled synchronously. If the system is time-invariant,

$$x_1 = \epsilon^{AT} x_0 + \int_0^T \epsilon^{A(T-\tau)} Bu(\tau) \, d\tau.$$

$$= \epsilon^{AT} x_0 + \left[ \int_0^T \epsilon^{A(T-\tau)} \, d\tau \right] Bu_0. \tag{9.77}$$

(Note that in taking the constant $Bu_0$ out of the integral, we must preserve the order of multiplication, since we are dealing with matrix and vector quantities.) The integral may now be evaluated as follows:

$$\int_0^T \epsilon^{A(T-\tau)} \, d\tau = \epsilon^{AT} \int_0^T \epsilon^{-A\tau} \, d\tau$$

$$= \epsilon^{AT} [-A^{-1} \epsilon^{-A\tau}]_0^T = \epsilon^{AT} [A^{-1}(I - \epsilon^{-AT})]. \tag{9.78}$$

Since $\epsilon^{AT}$ and $A^{-1}$ commute,

$$\int_0^T \epsilon^{A(T-\tau)} \, d\tau = A^{-1}(\epsilon^{AT} - I), \tag{9.79}$$

if $A^{-1}$ exists.

To find $x_2$ we may now simply treat $x_1$ as a new initial vector and write

$$x_2 = \epsilon^{AT} x_1 + \left[ \int_0^T \epsilon^{A(T-\tau)} \, d\tau \right] Bu_1, \tag{9.80}$$

or, in general,

$$x_{n+1} = \epsilon^{AT} x_n + \left[ \int_0^T \epsilon^{A(T-\tau)} \, d\tau \right] Bu_n. \tag{9.81}$$

Hence if we define

$$F \triangleq \epsilon^{AT}, \qquad G \triangleq \int_0^T \epsilon^{A(T-\tau)} \, d\tau B,$$

then our state equations become

$$x_{n+1} = Fx_n + Gu_n \tag{9.82}$$

$$y_n = Cx_n + Du_n, \tag{9.83}$$

where $C$ and $D$ are the same matrices as in the continuous equations. After $F$ and $G$ are computed, the computer simulation to generate the output $y_n$ from the input sequence $u_n$ involves only matrix multiplications, as indicated by equations 9.82 and 9.83.

A more accurate simulation algorithm results if we use a straight-line segment (trapezoidal) approximation to the continuous input signals of the

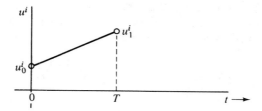

**Figure 9.7**   Straight-line Segment Spanning One Sampling Interval

system, as was done in Section 7.4. Let us denote the $i$th input function by $u^i(t)$ and approximate it over the first sampling interval by a straight-line segment, as indicated in Figure 9.7.

The expression for this straight-line segment is

$$u^i(t) = u_0^i + \frac{1}{T}(u_1^i - u_0^i)\, t. \tag{9.84}$$

In vector form,

$$u(t) = u_0 + \frac{1}{T}(u_1 - u_0)\, t = \left(1 - \frac{t}{T}\right)u_0 + \frac{t}{T}u_1, \tag{9.85}$$

where $u(t)$ is now a vector of straight-line segments, representing the approximation to the entire input vector over the first sampling interval. These expressions are valid for the interval $0 \leq t \leq T$. If the system starts at an initial state $x_0$, and we use our approximation for the input vector, then

$$
\begin{aligned}
x(T) \triangleq x_1 &= \epsilon^{AT}x_0 + \int_0^T \epsilon^{A(T-\tau)} Bu(\tau)\, d\tau \\
&= \epsilon^{AT}x_0 + \left[\int_0^T \epsilon^{A(T-\tau)}\, d\tau\right]Bu_0 - \left[\frac{1}{T}\int_0^T \tau\epsilon^{A(T-\tau)}\, d\tau\right]Bu_0 \\
&\quad + \left[\frac{1}{T}\int_0^T \tau\epsilon^{A(T-\tau)}\, d\tau\right]Bu_1.
\end{aligned}
\tag{9.86}
$$

Therefore, we may write

$$x_1 = Fx_0 + G_1 u_0 + G_2 u_1, \tag{9.87}$$

where

$$F \triangleq \epsilon^{AT} \tag{9.88}$$

$$G_1 \triangleq \int_0^T \epsilon^{A(T-\tau)}\, d\tau B - \frac{1}{T}\int_0^T \tau\epsilon^{A(T-\tau)}\, d\tau B \tag{9.89}$$

$$G_2 \triangleq \frac{1}{T}\int_0^T \tau\epsilon^{A(T-\tau)}\, d\tau B. \tag{9.90}$$

The terms $F$, $G_1$, and $G_2$ are just constant matrices, but equation 9.87 is not yet in our standard state form. However, we may easily put it in the form of a state difference equation by defining

$$U_0 \triangleq \begin{bmatrix} u_0 \\ u_1 \end{bmatrix} \quad \text{and} \quad G \triangleq [G_1 \quad G_2].$$

We then have

$$x_1 = Fx_0 + GU_0, \tag{9.91}$$

or, in general,

$$X_{n+1} = FX_n + GU_n. \tag{9.92}$$

Thus a pair of input sample vectors is used for each step in the iteration, since it requires a pair of input samples to determine each trapezoid. If $A^{-1}$ exists, then the integrals in the $G_1$ and $G_2$ matrices may be evaluated analytically, and one obtains (the details will be left as an exercise)

$$F = \epsilon^{AT} \tag{9.93}$$

$$G_1 = A^{-1}\left[(\epsilon^{AT} - I)\left(I - \frac{A^{-1}}{T}\right) + I\right]B \tag{9.94}$$

$$G_2 = A^{-1}\left[\frac{A^{-1}}{T}(\epsilon^{AT} - I) - I\right]B. \tag{9.95}$$

### State Difference Equations

Equation 9.82 is an example of a difference equation, and is used, together with the output equation 9.83, to describe the behavior of a discrete-time system. We arrived at this particular equation by starting with a continuous-time system and a sampled-data input vector. However, many systems do not operate in continuous time at all. Systems such as those involving birth-and-death processes, manufacture and inventory processes, probabilistic systems involving chains of events, and so forth, are by their very nature discrete. Hence, although the difference equation model may arise through the process of approximating the behavior of a continuous-time differential system, it may also arise because of the discrete-time nature of the system itself.

### Example 9.7

As an example of a system that is itself discrete, let us imagine a firm engaged in the business of assembling, testing, and delivering some item, as indicated in the flow diagram of Figure 9.8.

After assembly, each item is tested, and according to the outcome is either sent for delivery or returned for repair and retesting. After delivery, the items are again tested by the customer, and either accepted or returned for repair, as indicated. $A$ represents the number of items awaiting assembly,

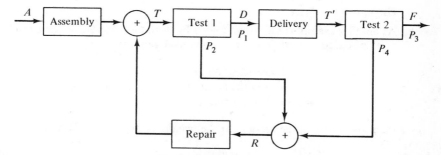

**Figure 9.8**   Example of a Discrete System

$T$ is the average number of items awaiting the first test, and so forth. Let us for simplicity assume that one unit of time is required for each of the operations (represented by the boxes) in Figure 9.8, which, in order to achieve a discrete approximation, will be assumed to occur synchronously in time. $P_1$ and $P_3$ represent the probabilities of rejection for the two tests, respectively. (We will assume that there is no limit to the number of items that can be processed simultaneously at each operation.)

Since any item is subjected to at least four delays in passing through the system, we would suspect that the system is of fourth order. In fact, we may write the following equations relating the input $A$, the output $F$, and the four variables $T$, $D$, $T'$, and $R$.

$$T_{n+1} = R_n + A_n$$

$$D_{n+1} = P_1 T_n$$

$$T'_{n+1} = D_n$$

$$R_{n+1} = P_2 T_n + P_4 T'_n \tag{9.96}$$

$$F_n = P_3 T'_{n-1} = \frac{P_3}{P_4} R_n - PT_{n-1}$$

$$= \frac{P_3}{P_4} R_n - \frac{P_2 P_3}{P_1 P_4} D_n$$

$$P_1 + P_2 = 1, \qquad P_3 + P_4 = 1.$$

In matrix form,

$$\begin{bmatrix} T_{n+1} \\ D_{n+1} \\ T'_{n+1} \\ R_{n+1} \end{bmatrix} \begin{bmatrix} 0 & 0 & 0 & 1 \\ P_1 & 0 & 0 & 0 \\ 0 & 1 & 0 & 0 \\ P_2 & 0 & P_4 & 0 \end{bmatrix} \begin{bmatrix} T_n \\ D_n \\ T'_n \\ R_n \end{bmatrix} + \begin{bmatrix} 1 \\ 0 \\ 0 \\ 0 \end{bmatrix} [A_n] \tag{9.97}$$

$$[F_n] = \begin{bmatrix} 0 & -\dfrac{P_2 P_3}{P_1 P_4} & 0 & \dfrac{P_3}{P_4} \end{bmatrix} \begin{bmatrix} T_n \\ D_n \\ T'_n \\ R_n \end{bmatrix}. \tag{9.98}$$

### Example 9.8

As another example of a system that is by its nature discrete, let us consider a savings account with interest compounded daily. If

$$C_n = \text{cash deposited on } n\text{th day}$$
$$X_n = \text{money in the account at beginning of } n\text{th day}$$
$$K = \text{daily interest rate,}$$

then the money in the account at the beginning of the $(n + 1)$th day is

$$X_{n+1} = X_n + KX_n + C_n = (K + 1)X_n + C_n. \qquad (9.99)$$

Thus the growth of money in the account is described by our standard difference equation.

### Solution of the State Difference Equation

We may write down a formal solution of equation 9.82 by using a simple process of iteration. If we start with an initial state $x_1$, we obtain

$$
\begin{aligned}
x_2 &= Fx_1 + Gu_1 \\
x_3 &= Fx_2 + Gu_2 = F^2x_1 + FGu_1 + Gu_2 \\
x_4 &= Fx_3 + Gu_3 = F^3x_1 + F^2Gu_1 + FGu_2 + Gu_3 \qquad (9.100) \\
&\qquad\qquad\qquad\quad . \\
&\qquad\qquad\qquad\quad . \\
&\qquad\qquad\qquad\quad .
\end{aligned}
$$

or

$$x_{n+1} = F^n x_1 + \sum_{i=1}^{n} F^{n-i} Gu_i. \qquad (9.101)$$

This is the discrete analog of the solution expressed by equation 9.66; the convolution integral of equation 9.66 is here replaced by the *convolution sum* in equation 9.88, and by setting $u_i = 0$ for all $i$, we have

$$x_{n+1} = F^n x_1, \qquad (9.102)$$

which is the solution of the homogeneous difference equation analogous to 9.65. The similarity between equations 9.66 and 9.88 should be noted by comparing the function of the integers $i$ and $n$ in the discrete solution to the time variables $t$ and $\tau$ in the continuous-time solution. The state transition matrix $\epsilon^{At}$ in the solution of the continuous system equations has as its counterpart the simple matrix power $F^n$ in the discrete system equations. Equations 9.88 and 9.89 may of course be very easily programmed for digital computer solution.

## PROBLEMS

**9.1** Express the following equations as first-order vector differential equations:
(a) $\ddot{x} + 2x = u(t)$.
(b) $\dddot{x} + 3\ddot{x} + 2\dot{x} + x = 0$.
(c) $\ddot{x} + \dot{x} + 2\theta = u(t)$
$\ddot{\theta} + \dot{\theta} + \dot{x} = 4x + 2u(t)$.

**9.2** Write the mesh equations for the circuit of Figure 9.9, and express them in state-variable form, using inductor currents and capacitor voltages as the state variables. Consider the two mesh currents to be the output variables.

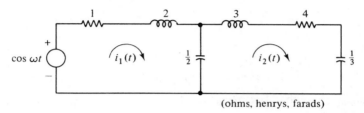

(ohms, henrys, farads)

**Figure 9.9**  Circuit of Problem 9.2

**9.3** Write the node-to-datum voltage equations for the circuit of Figure 9.10, and express them in state-variable form. Consider $v_2(t)$ to be the system output.

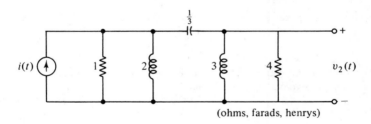

(ohms, farads, henrys)

**Figure 9.10**  Circuit of Problem 9.3

**9.4** Write the state equations for the point mass of Example 9.4, assuming that the mass moves in a viscous medium.

**9.5** Find $\epsilon^{At}$ for the following $A$ matrices:
(a) $A = \begin{bmatrix} 1 & 2 \\ 3 & 4 \end{bmatrix}$.
(b) $A = \begin{bmatrix} \frac{1}{2} & \frac{1}{4} \\ 1 & \frac{1}{2} \end{bmatrix}$.
(c) $A = \begin{bmatrix} 1 & 0 \\ 1 & 1 \end{bmatrix}$.

**9.6** Show that, if $x$ is a state vector and $x'$ is related to $x$ by

$$x' = Mx, \qquad x = M^{-1}x',$$

then $x'$ is also a state vector.

**9.7** (Digital computer exercise.) We wish to simulate the circuit of Figure 9.11 on a digital computer, so we may find the inductor current and the capacitor

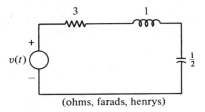

(ohms, farads, henrys)

**Figure 9.11**   Circuit of Problem 9.7

voltage for different types of voltage forcing functions. Assume that $v(t)$ is specified as

$$v(t) = \sin t\, u(t),$$

and approximate this by a piecewise constant function using a sampling interval of $T = 0.1$ s. Program the computer accordingly to generate the sampled state variables.

**9.8** (Digital computer exercise.) Assume that in the circuit of Figure 9.10,

$$i(t) = u(t).$$

Using a sampling interval of $T = 0.1$ s, program a digital computer to generate the sampled values of $v_2(t)$.

# A

# Complex Algebra
# and Euler's Formulas

## A.1 COMPLEX-NUMBER REPRESENTATIONS

A complex number consists of the sum of a real number and an imaginary number, and may be represented as

$$z = a + jb, \tag{A.1}$$

where

$$\text{Re}\,(z) = a \tag{A.2}$$

and

$$\text{Im}\,(z) = b. \tag{A.3}$$

(In this text we have used $j$ to denote $\sqrt{-1}$; hence $jb = \sqrt{-b^2}$.) We may also represent $z$ as a point in the complex plane, a rectangular coordinate system whose abscissa is $\text{Re}\,(z)$ and whose ordinate is $\text{Im}\,(z)$, as indicated in Figure A.1. Since the point $z$ uniquely determines a line segment from the origin to $z$, the complex number may also be specified by giving the length ($A$) of this segment (the magnitude or norm of the complex number, also denoted by $|z|$) and its angle ($\theta$) as measured from the positive real axis (also called the *argument* of the complex number and denoted by $\arg z$ or $\underline{/z}$). Thus, using these different representations, we may write

$$z = a + jb = |z|\ \underline{/\arg z}$$
$$= \sqrt{a^2 + b^2}\ \underline{\Big/\tan^{-1}\frac{b}{a}}. \tag{A.4}$$

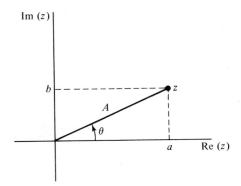

**Figure A.1**

From Figure A.1 it is also evident that

$$a = |z| \cos \theta = A \cos \theta$$
$$b = |z| \sin \theta = A \sin \theta,$$

(A.5)

so we may write

$$z = A \cos \theta + jA \sin \theta.$$

(A.6)

## A.2 EULER'S FORMULAS

If we consider $a$ and $b$ in equation A.1 to be variables, then $z$ is a complex variable, and we may define functions of $z$ which are analogous to functions of real variables. One of the most important functions is $\epsilon^z$, and it is defined by a power-series expansion analogous to that of the scalar exponential:

$$\epsilon^z = 1 + z + \frac{z^2}{2!} + \frac{z^3}{3!} + \ldots = \sum_{k=0}^{\infty} \frac{z^k}{k!}.$$

(A.7)

This series may be shown to converge (i.e., both real and imaginary parts converge) for $|z| < \infty$. If we replace $z$ by the complex variable $jz$ in this series, and make use of the fact that

$$j^2 = -1$$
$$j^3 = -j$$
$$j^4 = +1$$
$$j^5 = +j \qquad \text{etc.,}$$

we obtain

$$\epsilon^{jz} = \left(1 - \frac{z^2}{2!} + \frac{z^4}{4!} - \ldots\right) + j\left(z - \frac{z^3}{3!} + \frac{z^5}{5!} - \ldots\right).$$

(A.8)

The two series in equation A.8 however are just the series expansions for the functions $\cos z$ and $\sin z$, respectively; hence

$$\epsilon^{jz} = \cos z + j \sin z, \tag{A.9}$$

and similarly,

$$\epsilon^{-jz} = \cos z - j \sin z. \tag{A.10}$$

By adding equations A.9 and A.10, one may easily see that

$$\cos z = \tfrac{1}{2}(\epsilon^{jz} + \epsilon^{-jz}), \tag{A.11}$$

and by subtracting the same equations one obtains

$$\sin z = \frac{1}{2j}(\epsilon^{jz} - \epsilon^{-jz}). \tag{A.12}$$

Equations A.9 through A.12 are commonly referred to as *Euler's formulas*.

If we let $z$ be the real variable $\theta$ in Euler's formulas, then equation A.6 may be written

$$z = A \cos \theta + jA \sin \theta = A\epsilon^{j\theta}, \tag{A.13}$$

where $A$ and $\theta$ are the magnitude and argument of $z$, respectively, and this is yet another representation for the complex number $z$.

## A.3 ARITHMETIC OPERATIONS WITH COMPLEX NUMBERS

Addition and subtraction of complex numbers are most easily performed using the rectangular representation, while multiplication and division are most easily carried out in polar or exponential form, as indicated:

$$\begin{aligned} z_1 + z_2 &= (a_1 + jb_1) + (a_2 + jb_2) \\ &= (a_1 + a_2) + j(b_1 + b_2). \end{aligned} \tag{A.14}$$

$$\begin{aligned} z_1 - z_2 &= (a_1 + jb_1) - (a_2 + jb_2) \\ &= (a_1 - a_2) + j(b_1 - b_2). \end{aligned} \tag{A.15}$$

$$z_1 \cdot z_2 = A_1\epsilon^{j\theta_1} \cdot A_2\epsilon^{j\theta_2} = A_1 A_2^{j(\theta_1 + \theta_2)}. \tag{A.16}$$

$$z_1 \div z_2 = \frac{A_1}{A_2}\epsilon^{j(\theta_1 - \theta_2)}. \tag{A.17}$$

Once an answer is obtained in either rectangular or exponential form, a conversion to the other form may be easily made using the relationships of equations A.4 and A.13.

If one has the ratio of two complex numbers in rectangular form, the real and imaginary parts of the ratio may easily be obtained by multiplying both numerator and denominator by the complex conjugate of the denominator (this time we will multiply in rectangular coordinates):

$$\frac{a_1 + jb_1}{a_2 + jb_2} = \frac{a_1 + jb_1}{a_2 + jb_2} \frac{a_2 - jb_2}{a_2 - jb_2}$$

$$= \frac{a_1 a_2 + b_1 b_2}{a_2^2 + b_2^2} + j\frac{a_2 b_1 - a_1 b_2}{a_2^2 + b_2^2}. \tag{A.18}$$

This operation is called *rationalization* of a complex fraction.

## A.4 ROOTS, POWERS, AND LOGARITHMS OF COMPLEX NUMBERS

These operations are most easily performed using an exponential representation for the complex numbers involved. For example,

$$z^n = (A\epsilon^{j\theta})^n = A^n \epsilon^{jn\theta}. \tag{A.19}$$

$$\sqrt[n]{z} = z^{1/n} = A^{1/n} \epsilon^{j\theta/n}$$

$$= A^{1/n} \epsilon^{j(\theta + k2\pi)/n} \tag{A.20}$$

where $k$ is any integer. (The value for $k = 0$ is called the *principal value*.)

### Example A.1

Find the three cube roots of $-1$.

$$-1 = 1\epsilon^{j180°}$$

$$\sqrt[3]{-1} = \sqrt[3]{1}\, \epsilon^{j(180°+k360°)/3} = 1\epsilon^{j(60°+k120°)}. \tag{A.21}$$

Thus the three cube roots of $-1$ are

$$1\epsilon^{j60°}, \qquad 1\epsilon^{j180°}, \qquad 1\epsilon^{j(-60°)}.$$

The natural logarithm of a complex number may be found as follows:

$$\ln z = \ln A\epsilon^{j\theta} = \ln A + j\theta$$

$$= \ln A + j(\theta + k2\pi). \tag{A.22}$$

(Again, the value for $k = 0$ is called the principal value of the logarithm.)

## PROBLEMS

**A.1** Express the following numbers in "rectangular" form:
(a) $5\epsilon^{j(\pi/2)}$     (b) $2\epsilon^{j30°}$
(c) $\epsilon^{j225°}$     (d) $3\underline{/60°}$

**A.2** Express the following numbers in exponential form:
(a) $1 + j2$     (b) $-2 + j3$
(c) $-1 - j1$     (d) $1 - j2$

**A.3** Show the numbers of Problems A.1 and A.2 as points in the complex plane.

**A.4** Rationalize the following fractions:

(a) $\dfrac{1 + j3}{2 - j4}$     (b) $\dfrac{1 - j1}{-2 + j2}$

**A.5** Given the following two complex numbers:

$$z_1 = 1 + j2 \quad \text{and} \quad z_2 = 2 - j3,$$

find the following (in any form):
(a) $z_1 + z_2$     (b) $z_1 z_2$
(c) $z_1/z_2$     (d) $z_1^4$
(e) $\sqrt[5]{z_2}$     (f) $\ln z_1$

**A.6** Find the six sixth roots of unity, and indicate their position in the complex plane.

# B

# Solution of Simultaneous Equations Using Matrices[1] and Determinants

## B.1 CRAMER'S RULE

Consider a set of $n$ linear algebraic equations in $n$ unknowns, $x_1, x_2, \ldots, x_n$:

$$a_{11}x_1 + a_{12}x_2 + \ldots + a_{1n}x_n = b_1$$
$$a_{21}x_1 + a_{22}x_2 + \ldots + a_{2n}x_n = b_2 \qquad (B.1)$$
$$\cdot$$
$$\cdot$$
$$\cdot$$
$$a_{n1}x_1 + a_{n2}x_2 + \ldots + a_{nn}x_n = b_n,$$

or, in matrix form,

$$Ax = b. \qquad (B.2)$$

We will let $\Delta$ be the determinant of the $A$ matrix; i.e.,

$$\Delta = \begin{vmatrix} a_{11} & a_{12} & \cdots & a_{1n} \\ a_{21} & a_{22} & & a_{2n} \\ \cdot & & & \cdot \\ \cdot & & & \cdot \\ \cdot & & & \cdot \\ a_{n1} & a_{n2} & \cdots & a_{nn} \end{vmatrix}. \qquad (B.3)$$

[1]It is assumed that the reader has some prior familiarity with matrix representations and elementary operations (addition and multiplication) with matrices and vectors. Certain other selected aspects of matrix analysis (e.g., properties of the matrix exponential) are discussed in Chapter 9.

If $\Delta \neq 0$, then the unique solution for the $k$th unknown variable $x_k$ is

$$x_k = \frac{\Delta_k}{\Delta}, \tag{B.4}$$

where $\Delta_k$ is the determinant of the $A$ matrix with its $k$th column replaced by the column of numbers on the right-hand side of the equation; i.e., $b_1, b_2, \ldots, b_n$. This is known as *Cramer's rule*.

**Example B.1**

If

$$a_{11}x_1 + a_{12}x_2 = b_1$$
$$a_{21}x_1 + a_{22}x_2 = b_2, \tag{B.5}$$

then

$$x_1 = \frac{\begin{vmatrix} b_1 & a_{12} \\ b_2 & a_{22} \end{vmatrix}}{\begin{vmatrix} a_{11} & a_{12} \\ a_{21} & a_{22} \end{vmatrix}} = \frac{b_1 a_{22} - b_2 a_{12}}{a_{11}a_{22} - a_{12}a_{21}} \tag{B.6}$$

$$x_2 = \frac{\begin{vmatrix} a_{11} & b_1 \\ a_{21} & b_2 \end{vmatrix}}{\begin{vmatrix} a_{11} & a_{12} \\ a_{21} & a_{22} \end{vmatrix}} = \frac{b_2 a_{11} - b_1 a_{21}}{a_{11}a_{22} - a_{12}a_{21}}. \tag{B.7}$$

## B.2 GENERAL THEORY OF SIMULTANEOUS EQUATIONS

### Rank of a Matrix

If a matrix is not square, or if it has a determinant that is zero, then it is useful to define the rank of the matrix. The *rank* of a matrix is the largest number $r$ such that at least one $r$th-order determinant different from zero can be formed from the matrix by deleting rows and/or columns from the original matrix. If a square matrix has a nonzero determinant, then of course the rank is the same as the order of the matrix.

### Linear Independence

Consider a set of $m$ functions

$$f_i(x_1, x_2, \ldots, x_n) \qquad (i = 1, 2, \ldots, m).$$

This set of functions is said to be *linearly independent* if none of them can be expressed as linear combinations of the others; i.e., if and only if

$$\sum_{i=1}^{m} C_i f_i = 0 \Longrightarrow C_1 = C_2 = \ldots = C_m = 0. \tag{B.8}$$

Otherwise, the functions are said to be linearly dependent:

$n$ linear functions $\qquad \displaystyle\sum_{k=1}^{n} a_{ik} x_k \qquad (i = 1, 2, n)$

are linearly independent if and only if the determinant of the matrix of coefficients $a_{ik}$ is nonzero; i.e., det $[a_{ik}] \neq 0$.

$m$ linear functions $\qquad \displaystyle\sum_{k=1}^{n} a_{ik} x_k \qquad (i = 1, 2, \ldots, m)$

are linearly independent if and only if the $(m \times n)$ matrix $[a_{ik}]$ is of rank $m$.

### General Simultaneous Linear Equations

Consider again the equation $Ax = b$, where the $A$ matrix is now $(m \times n)$. It may be shown that this system of equations ($m$ equations in $n$ unknowns) possesses a solution if and only if the matrices

$$
\begin{bmatrix} a_{11} & a_{12} & \cdots & a_{1n} \\ \cdot & & & \\ \cdot & & & \\ \cdot & & & \\ a_{m1} & \cdots\cdots & & a_{mn} \end{bmatrix}
\quad \text{and} \quad
\begin{bmatrix} a_{11} & a_{12} & \cdots & a_{1n} & b_1 \\ \cdot & & & & \\ \cdot & & & & \\ \cdot & & & & \\ a_{m1} & \cdots\cdots & & a_{mn} & b_m \end{bmatrix}
$$

have the *same rank*. Otherwise, the equations are *inconsistent*, and no solution exists.

If $m = n = r$ (the rank), then the equations possess a unique solution which may be found using Cramer's rule. On the other hand, if both of the above matrices are of rank $r < m$, then the equations are *linearly dependent*. One may then solve the subset of $r$ linearly independent equations, and these solutions will also satisfy the remaining $m - r$ equations. The $r$ independent equations determine $r$ of the unknowns as linear functions of the remaining $n - r$ unknowns, which remain arbitrary. Thus an infinite number of specific solutions to the set of equations exists.

### $n$ Homogeneous Equations in $n$ Unknowns

If we apply Cramer's rule to the set of equations $Ax = 0$, then $\Delta_k = 0$ for all $k$, since $\Delta_k$ will be the determinant of a matrix that has a column of zeros. The set of equations will, of course, always have the trivial solution $x_1 = x_2 = \ldots = x_n = 0$. For a nontrivial solution, however, we must have det $[a_{ij}] = \Delta = 0$, and in this case there are exactly $n - r$ linearly independent solution sets, where $r$ is the rank of $\Delta$. If we denote each of these solution sets by the superscript $j$, then the most general form of the solution may be written as

$$x_i = \sum_{j=1}^{n-r} C_j x_i^{(j)}, \tag{B.9}$$

where the $C_j$ are arbitrary constants.

## B.3  MATRIX INVERSION

Consider again equation B.2, where $A$ is square and $\Delta \neq 0$. Then we may write

$$x = A^{-1}b, \tag{B.10}$$

where $A^{-1}$ is the inverse of the $A$ matrix. This inverse may be obtained as follows:

1. Find $\Delta = \det [a_{ij}]$.
2. Find $A^T$ ($A$ transpose). (The transpose of a matrix is obtained by interchanging its rows and columns.)
3. Find the matrix $Cf[A^T]$ whose elements are the cofactors of the elements of $A^T$. [The cofactor of an element $a_{ij}$ is equal to $(-1)^{i+j}$ times the determinant formed by deleting the $i$th row and $j$th column of the original matrix.]
4. $A^{-1} = (1/\Delta) [Cf(A^T)]$.

## PROBLEMS

**B.1**  Solve the following using Cramer's rule:

(a) $x_1 + x_2 = 1$
$3x_1 + 2x_2 = 4$

(b) $3x_1 - x_2 + x_3 = 0$
$4x_1 + 2x_2 - x_3 = 1$
$5x_1 + x_2 + x_3 = 2$

**B.2**  Find solutions to the following:

(a) $x_1 + x_2 + x_3 = 1$
$2x_1 + 2x_2 + 2x_3 = 2$
$2x_1 - x_2 - x_3 = 3$

(b) $x_1 + x_2 = 2$
$3x_1 + 3x_2 = 6$

**B.3**  Invert the following matrices:

(a) $A = \begin{bmatrix} 0 & 1 \\ -2 & -3 \end{bmatrix}$

(b) $A = \begin{bmatrix} 1 & 2 & 3 \\ 1 & -1 & 0 \\ 4 & 5 & 0 \end{bmatrix}$

# C

# Problems
# Similar to Selected Ones
# in the Text,
# with Solutions

The problems solved in this appendix are closely related to certain problems appearing at the ends of the various chapters; the correspondence is indicated by the numbering (e.g., Problem C1.6 is closely related to Problem 1.6).

### CHAPTER 1

**C1.5**

A two-dimensional vector field is specified by

$$\bar{V} = x^2 \bar{i}_y$$

Find the line integral along the straight-line segment connecting the points $(0, 0)$ and $(\sqrt{3}, 1)$.

**Solution:**
Along the line segment, $dl = dx/\cos 30° = 2dx/\sqrt{3}$.

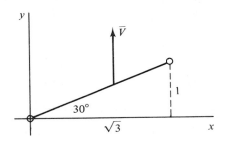

The component of $\bar{V}$ in the direction of $\overline{dl}$ is

$$V_{dl} = V_y \cos 60° = 0.5x^2.$$

Therefore, integrating with respect to $x$,

$$\int_{(0,0)}^{(\sqrt{3},1)} \bar{V} \cdot dl = \int_0^{\sqrt{3}} (0.5x^2)\left(\frac{2}{\sqrt{3}} dx\right) = \frac{1}{\sqrt{3}} \int_0^{\sqrt{3}} x^2\, dx$$

$$= \frac{1}{\sqrt{3}}\left[\frac{1}{3}x^3\right]_0^{\sqrt{3}} = \frac{(\sqrt{3})^3}{3 \cdot \sqrt{3}} = 1.$$

### C1.6

A three-dimensional uniform field is specified by

$$\bar{V} = V_0 \bar{i}_x$$

Find the surface integral over the plane surface bounded by the four straight-line segments meeting at the points $(1, 0, 0)$, $(0, 0, 1)$, $(1, 1, 0)$, and $(0, 1, 1)$.

**Solution:**

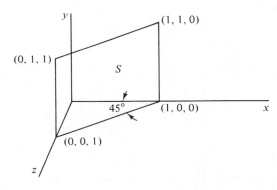

$$ds = \frac{dy\, dz}{\cos 45°} = \sqrt{2}\, dy\, dz.$$

The component of $\bar{V}$ in the direction of $\overline{ds}$ is

$$V_{ds} = V_0 \cos 45° = \frac{V_0}{\sqrt{2}}.$$

Therefore, integrating with respect to $y$ and $z$,

$$\int_s \bar{V} \cdot \overline{ds} = \int_0^1 \int_0^1 \frac{V_0}{\sqrt{2}} (\sqrt{2}\, dy\, dz) = V_0 \int_0^1 \int_0^1 dy\, dz = V_0.$$

### C1.7

Find an expression for $\bar{i}_x \cdot \overline{ds}$ over the surface of a sphere of unit radius, in spherical coordinates.

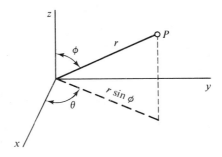

**Solution:**

A spherical coordinate representation for a point $P$ is shown in the figure. The element of surface area over a sphere of constant radius is

$$ds = (r \, d\theta \sin \phi)(r \, d\phi).$$

Therefore,

$$\overline{ds} = r^2 \sin \phi \, d\theta \, d\phi \, \bar{i}_r.$$

Also, from the geometry,

$$\bar{i}_x \cdot \overline{ds} = (r^2 \sin \phi \, d\theta \, d\phi) \bar{i}_x \cdot \bar{i}_r = (r^2 \sin \phi \, d\theta \, d\phi)(\sin \phi \cos \theta).$$

Therefore, over a sphere of unit radius,

$$\bar{i}_x \cdot \overline{ds} = \sin^2 \phi \cos \theta \, d\theta \, d\phi.$$

### C1.15

A uniform surface charge of $\rho_a$ coulombs per square meter is distributed on a plane surface of infinite extent. Use Gauss's law to find the electric field everywhere.

**Solution:**

Choose a Gaussian cylinder, the ends of which are of unit area parallel to the plane, and the sides of which are normal to the plane. From symmetry, the electric field is everywhere normal to the plane. Therefore, applying Gauss's law, we obtain

$$\int_s \bar{E} \cdot \overline{ds} = 2E_y \, (\text{unit area}) = \frac{\rho_a \, (\text{unit area})}{\epsilon}.$$

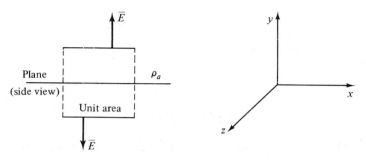

Therefore,
$$E_y = \frac{\rho_a}{2\epsilon}.$$

(Compare this result with Example 1.4.)

### *C1.21*

A fixed charge of $q$ coulombs is put on the plates of a parallel-plate capacitor of area $A$ and plate separation distance $x$ in free space ($\epsilon_0$). Find the force between the two plates.

### *Solution:*

Consider a small virtual displacement ($dx$) of one of the plates, as indicated in the figure. This displacement $dx$ will cause a change $dw$ in the stored energy, and

$$F = -\frac{dw}{dx}$$

But
$$w(x) = \frac{1}{2} C(x) v^2(x) = \frac{1}{2} C(x) \left[\frac{q}{C(x)}\right]^2 = \frac{1}{2}\frac{q^2}{C(x)} = \frac{1}{2}\frac{q^2}{\epsilon A} x$$

Therefore,
$$F = -\frac{dw}{dx} = -\frac{1}{2}\frac{q^2}{\epsilon A}.$$

(The negative sign indicates a force opposite in direction to $dx$, or a force of attraction between the plates.)

## *CHAPTER 2*

### *C2.4*

The current in a wire is given by

$$i(t) = \begin{cases} \cos \omega t & (t > 0) \\ 0 & (t < 0). \end{cases}$$

Find the charge that has passed through any cross section of the wire at $t = 1$ s.

**Solution:**

$$q(1) = \int_0^1 \cos \omega t \, dt = \frac{1}{\omega} \sin \omega t \bigg|_0^1 = \frac{\sin \omega}{\omega} \quad \text{coulombs.}$$

[What is the difficulty in answering this question if $i(t) = \cos \omega t$ for *all* time?]

### C2.9

Show that a circular orbit is one possible path for an electron traveling at a constant speed through a uniform $\bar{B}$ field.

**Solution:**

The electron may travel in a circular orbit of radius $R$ in a plane perpendicular to the $\bar{B}$ field. If the constant tangential speed is $u_0$ and the charge on the electron is denoted by $q_e$, then the centripedal force on the electron, directed toward the center of the orbit, is

$$F = \frac{mu_0^2}{R} = q_e u_0 B, \qquad \text{giving } R = \frac{mu_0}{q_e B}.$$

### C2.10

Write an expression for the integral required in Problem 2.10.

**Solution:**

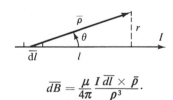

$$d\bar{B} = \frac{\mu}{4\pi} \frac{I \, \overline{dl} \times \bar{\rho}}{\rho^3}.$$

Using the geometry of the figure, the magnitude of $d\bar{B}$ may be expressed as

$$dB = \frac{\mu I}{4\pi} \frac{\rho \, dl \sin \theta}{\rho^3} = \frac{\mu I}{4\pi} \frac{\sin \theta}{\rho^2} \, dl$$

If we choose to integrate with respect to $\theta$, we may write

$$l = r \cot \theta$$

$$dl = r \, d(\cot \theta) = -\frac{r \, d\theta}{\sin^2 \theta}$$

$$\rho = \frac{r}{\sin \theta}.$$

Substituting these into the expression for *dB* and integrating, we get

$$B = -\frac{\mu I}{4\pi r} \int_0^\pi \sin\theta \, d\theta.$$

### C2.16

A straight wire carries a current of *I* amperes. Find the flux of *B* through a 3- × 2-m rectangle oriented as shown in the figure.

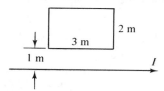

*Solution:*

From Problem 2.10, $B = \mu I/2\pi r$ and is perpendicular to the plane of the rectangle. Therefore,

$$\phi = \int_s \bar{B} \cdot \overline{ds} = 3 \int_1^3 \frac{\mu I}{2\pi r} \, dr = \frac{3\mu I}{2\pi} [\ln r]_1^3 = 0.5245\mu I.$$

### C2.18

Write an expression for the integral required in Problem 2.18.

*Solution:*

By considering current elements in diametrically opposite pairs, it is clear that the components of $\bar{B}$ parallel to the plane of the loop cancel, leaving only a normal component (in the *z* direction).

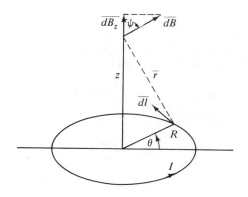

$$\overline{dB} = \frac{\mu}{4\pi} \frac{I \, \overline{dl} \times \bar{r}}{r^3}.$$

(Note that for all points on the axis of the loop, $\overline{dl} \perp \bar{r}$.)

$$dB = \frac{\mu I \, dl}{4\pi \, r^2} \quad \text{and} \quad dB_z = dB \cos \psi = \frac{\mu I \, dl}{4\pi \, r^2} \cos \psi.$$

If we choose to integrate with respect to $\theta$, we may write

$$dl = R \, d\theta, \qquad r^2 = R^2 + z^2, \qquad \cos \psi = \frac{R}{(R^2 + z^2)^{1/2}}.$$

Therefore,

$$B_z = \frac{\mu I R^2}{4\pi (R^2 + z^2)^{3/2}} \int_0^{2\pi} d\theta.$$

### C2.20

A circular loop of wire of radius $R$ rotates at a constant angular velocity of $\omega$ radians/second in a uniform $\bar{B}$ field, as indicated in the figure. Find the total EMF induced around the loop.

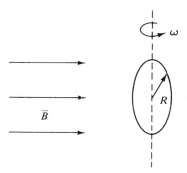

### Solution:

If we assume that the plane of the loop is perpendicular to the field at $t = 0$, the flux of $\bar{B}$ through the loop will be

$$\psi = \pi R^2 \cos \omega t,$$

and, from Faraday's law,

$$\oint \bar{E} \cdot \overline{dl} = -\frac{d\psi}{dt} = \pi R^2 \omega \sin \omega t.$$

**C2.21**

Find the force per unit length on a straight wire carrying a current $I$ amperes perpendicular to a uniform $\bar{B}$ field.

**Solution:**

$\overline{dF} = I\,\overline{dl} \times \bar{B}$; therefore, $F = IB$ newtons/meter, and this force acts in a direction normal to both the field and the wire.

## CHAPTER 3

**C3.1**

Sketch the integral and derivative functions for the time function $f(t)$.

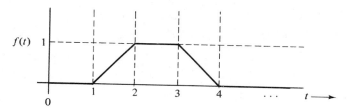

**Solution:**

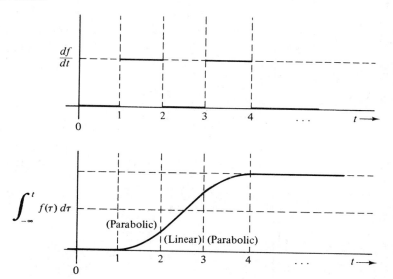

**C3.3**

A mechanical system consists of a mass and spring, with viscous damping, as indicated in the figure. The mass is subject to an external force $F(t)$.

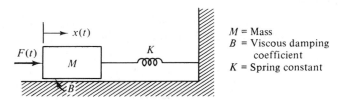

$M$ = Mass
$B$ = Viscous damping coefficient
$K$ = Spring constant

Find analogous electrical circuits, with (a) force analogous to voltage, and (b) force analogous to current. $x(t)$ represents the displacement of the mass.

***Solution:***

Equating the forces acting on the mass, we obtain

$$F(t) = M\ddot{x} + B\dot{x} + Kx,$$

or in terms of the velocity $V$,

$$F(t) = M\dot{V} + BV + K \int_{-\infty}^{t} V(\tau)\, d\tau.$$

(a) By applying Kirchhoff's voltage law to the series circuit shown, the same form of equation is obtained with the analogous quantities as indicated:

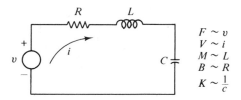

$F \sim v$
$V \sim i$
$M \sim L$
$B \sim R$
$K \sim \dfrac{1}{c}$

(b) If Kirchhoff's current law is applied to the parallel circuit shown, the same equation is obtained with the following analogies:

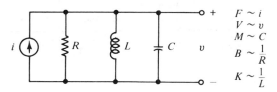

$F \sim i$
$V \sim v$
$M \sim C$
$B \sim \dfrac{1}{R}$
$K \sim \dfrac{1}{L}$

### C3.9

Write equations for the circuit shown, in terms of (a) node-to-datum voltages and (b) mesh currents.

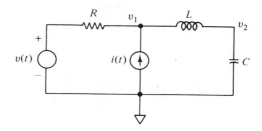

**Solution:**

(a) $\dfrac{v - v_1}{R} + i(t) + \dfrac{1}{L} \displaystyle\int_{-\infty}^{t} (v_2 - v_1)\, d\tau = 0$

$\dfrac{1}{L} \displaystyle\int_{-\infty}^{t} (v_1 - v_2)\, d\tau - C\dfrac{dv_2}{dt} = 0.$

(b) First redraw the circuit as follows:

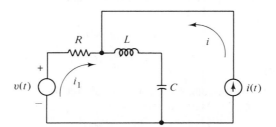

There is only one *unknown* mesh current ($i_1$) and its equation is

$$v(t) - i_1 R - L\frac{d}{dt}(i_1 + i) - \frac{1}{c}\int_{-\infty}^{t} (i_1 + i)\, d\tau = 0.$$

### C3.16

In the circuit shown, solve for $I_2$ using determinants.

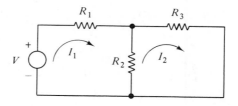

*Solution:*

Using Kirchhoff's voltage law,

$$V - I_1 R_1 - R_2(I_1 - I_2) = 0$$
$$-R_2(I_2 - I_1) - R_3 I_2 = 0.$$

These may be rewritten as

$$(R_1 + R_2)I_1 - R_2 I_2 = V$$
$$-R_2 I_1 + (R_2 + R_3)I_2 = 0.$$

Hence, by Cramer's rule,

$$I_2 = \frac{\begin{vmatrix} R_1 + R_2 & V \\ -R_2 & 0 \end{vmatrix}}{\begin{vmatrix} R_1 + R_2 & -R_2 \\ -R_2 & R_2 + R_3 \end{vmatrix}} = \frac{R_2 V}{R_1 R_2 + R_1 R_3 + R_2 R_3}.$$

## C3.18

Determine whether or not the following equation is linear:

$$t^2 \frac{dx}{dt} + 3tx(t) = y(t).$$

*Solution:*

Let $x_1(t)$ be the solution for the forcing function $y_1(t)$, and let $x_2(t)$ be the solution for the forcing function $y_2(t)$.
Then

$$t^2 \frac{dx_1}{dt} + 3tx_1 = y_1$$

$$t^2 \frac{dx_2}{dt} + 3tx_2 = y_2.$$

If we multiply the one equation by an arbitrary constant ($a$) and the other by ($b$) and add the two equations, we obtain

$$at^2 \frac{dx_1}{dt} + 3atx_1 + bt^2 \frac{dx_2}{dt} + 3btx_2 = ay_1 + by_2,$$

or

$$t^2 \frac{d}{dt}(ax_1 + bx_2) + 3t(ax_1 + bx_2) = ay_1 + by_2.$$

Hence $(ax_1 + bx_2)$ is the solution for the forcing function $(ay_1 + by_2)$, and the equation is linear.

## CHAPTER 4

### C4.5

For the circuit shown, find $\dot{v}(0+)$ and $\ddot{v}(0+)$, if $v(0+) = 3$ V and $i_L(0+) = 2$ A.

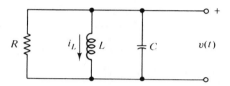

### Solution:

The equation for $v(t)$ is

$$\frac{v}{R} + C\frac{dv}{dt} + \frac{1}{L}\int_{-\infty}^{t} v(\tau)\,d\tau = 0.$$

Evaluated at $t = 0+$, the equation is

$$\frac{v(0+)}{R} + C\dot{v}(0+) + i_L(0+) = 0,$$

and with the numbers given,

$$\frac{3}{R} + C\dot{v}(0+) + 2 = 0, \qquad \text{which gives } \dot{v}(0+) = \frac{-(3 + 2R)}{RC}.$$

To find $\ddot{v}(0+)$ we must differentiate the equation for $v(t)$, and then evaluate it at $t = 0+$:

$$\frac{\dot{v}(0+)}{R} + C\ddot{v}(0+) + \frac{v(0+)}{L} = 0$$

$$\frac{-(3 + 2R)}{R^2C} + C\ddot{v}(0+) + \frac{3}{L} = 0$$

$$\ddot{v}(0+) = \frac{3L + 2RL - 3R^2C}{R^2C^2L}.$$

### C4.9

In the circuit shown, the switch is thrown in the direction indicated at $t = 0$. Find $V_{L_1}(0+)$ and $V_{L_2}(0+)$.

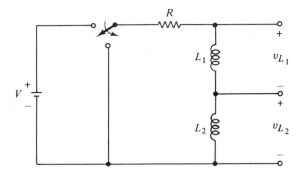

**Solution:**

At $t = 0+$ the equivalent circuit is

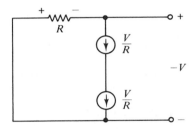

The question is: How does the voltage $(-v)$ divide between the two inductors (equivalent current sources)? Since the current (and its derivative) is the same in the two inductors, and since

$$V_{L_1}(0+) + V_{L_2}(0+) = -V$$
$$L_1 i(0+) + L_2 i(0+) = -V,$$

the initial voltage across each inductor is directly proportional to its inductance, and

$$V_{L_1}(0+) = \frac{-VL_1}{L_1 + L_2}, \qquad V_{L_2}(0+) = \frac{-VL_2}{L_1 + L_2}.$$

## CHAPTER 5

**C5.1**

In the following equation, $f(t)$ is sinusoidal, with unit peak value and zero phase reference. Find the peak value and phase angle of $x(t)$:

$$\frac{dx}{dt} + 2x = \frac{df}{dt} + f.$$

**Solution:**

$$H(j\omega) = \frac{j\omega + 1}{j\omega + 2}$$

$$|x| = H(j\omega)|\cdot 1 = \frac{|j\omega + 1|}{|j\omega + 2|} = \sqrt{\frac{\omega^2 + 1}{\omega^2 + 4}}$$

$$\underline{/x} = \underline{/H(j\omega)} + 0 = \underline{/j\omega + 1} - \underline{/j\omega + 2} = \tan^{-1}\omega - \tan^{-1}\frac{\omega}{2}.$$

### C5.3

Find the impedance, admittance, reactance, and susceptance for the circuit shown, at a frequency of $\omega = 1$ rad/s.

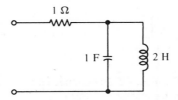

**Solution:**

In terms of impedances, the circuit is

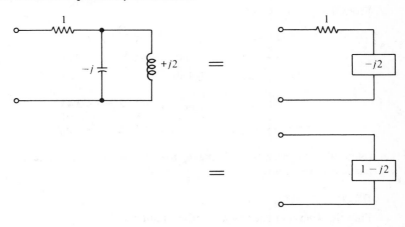

$$Z = 1 - 2j$$

$$Y = \frac{1}{1 - 2j}\frac{1 + 2j}{1 + 2j} = \frac{1 + 2j}{5}; \qquad X = -2, \qquad B = \frac{2}{5}.$$

### C5.8

Find the average and RMS value of the periodic waveform shown.

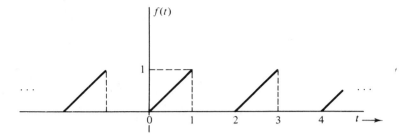

**Solution:**

$$f_{av}(t) = \frac{1}{2} \int_0^2 f(t)\, dt = \frac{1}{2} \int_0^1 t\, dt = \frac{1}{4} .$$

$$f_{RMS}(t) = \left[ \frac{1}{2} \int_0^2 f^2(t)\, dt \right]^{1/2} = \frac{1}{2} \left[ \int_0^1 t^2\, dt \right]^{1/2} = \frac{1}{\sqrt{6}} = 0.4082.$$

### C5.9

The circuit of Problem C5.3 is connected to a sinusoidal voltage source of 2 V RMS and a frequency of 1 rad/s. Find the complex power supplied by the source, the real power supplied by the source, and the power factor.

**Solution:**

From the solution of C5.3,

$$Y = 0.2 + j0.4, \qquad Y^* = 0.2 - j0.4$$
$$S = V_{RMS}^2 Y^* = 4(0.2 - j0.4) = 0.8 - j1.6$$
$$P = \text{Re}\,(S) = 0.8\ \text{W}$$
$$\theta = \tan^{-1}\left( -\frac{1.6}{0.8} \right) = -\tan^{-1} 2 = -63.43°$$

power factor $= \cos \theta = 0.4472.$

(Note that $\theta$ is negative; this means that the circuit looks capacitive at this particular frequency.)

### C5.18

Find the Thévenin equivalent for the circuit shown.

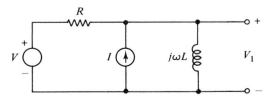

**Solution:**

The equation for the open-circuited voltage ($V_1$) which appears across the output terminals is (from Kirchhoff's current law)

$$\frac{V - V_1}{R} + I - \frac{V_1}{j\omega L} = 0.$$

Thus the Thévenin equivalent voltage source is

$$V_{\text{Th}} = V_1 = \left(\frac{V}{R} + I\right)\frac{j\omega RL}{R + j\omega L}.$$

If we short out the voltage source and open-circuit the current source, then the impedance seen looking back into the circuit from the output terminals is

$$Z_{\text{Th}} = \frac{j\omega RL}{R + j\omega L}.$$

**C5.20**

Find the Thévenin equivalent voltage ($V_1$) in the circuit of Problem C5.18 by using superposition.

**Solution:**

The output voltage ($V_{11}$) due to the voltage source alone (with the current source open-circuited) is

$$V_{11} = \frac{j\omega LV}{R + j\omega L}$$

(a voltage divider).
The output voltage ($V_{12}$) due to the current source alone (with the voltage source short-circuited) is

$$V_{12} = I\frac{j\omega RL}{R + j\omega L}.$$

Therefore, by superposition,

$$V_1 = V_{11} + V_{12} = \left(\frac{V}{R} + I\right)\frac{j\omega RL}{R + j\omega L}.$$

## CHAPTER 6

**C6.4**

Find the Fourier series of the halfwave-rectified cosine wave shown in the figure.

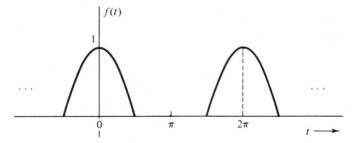

**Solution:**

Since $f(t)$ is an even function, the series contains only cosine terms.

$$T = 2\pi, \qquad \omega_0 = \frac{2\pi}{T} = 1$$

$$a_0 = \frac{1}{T} \int_T f(t)\, dt = \frac{1}{2\pi} \int_{-\pi/2}^{+\pi/2} \cos t\, dt = \frac{1}{\pi}$$

$$a_1 = \frac{1}{\pi} \int_{-\pi/2}^{+\pi/2} \cos^2 t\, dt = \frac{1}{\pi} \cdot \frac{\pi}{2} = \frac{1}{2}$$

For $n > 1$,

$$a_n = \frac{1}{\pi} \int_{-\pi/2}^{+\pi/2} \cos t \cos nt\, dt = \frac{1}{\pi} \left[ \frac{\sin (1 - n)t}{2(1 - n)} + \frac{\sin (1 + n)t}{2(1 + n)} \right]_{-\pi/2}^{+\pi/2}$$

$$= \frac{1}{\pi} \left[ \frac{\sin (1 - n)\pi/2}{1 - n} + \frac{\sin (1 + n)\pi/2}{1 + n} \right].$$

Thus, $a_2 = 2/3\pi$, $a_3 = 0$, $a_4 = -2/15\pi$, $a_5 = 0$, $a_6 = 2/35\pi$, etc., or

$$a_n = \frac{(-1)^{(n/2)+1}}{\pi} \frac{2}{n^2 - 1}, \qquad n = 2, 4, 6, \ldots$$

$$= 0, \qquad\qquad n = 3, 5, 7, \ldots.$$

### C6.7, C6.12

What is the power spectrum of the halfwave-rectified sinusoid of Problem C6.4?

**Solution:**

$$\alpha_0 = a_0 = \frac{1}{\pi}$$

$$\alpha_n = \frac{a_n - jb_n}{2}, \qquad\qquad n > 0$$

$$\alpha_1 = \frac{a_1}{2} = \frac{1}{4}$$

$$\alpha_n = \frac{a_n}{2} = \frac{(-1)^{(n/2)+1}}{\pi} \frac{1}{n^2 - 1}, \qquad n = 2, 4, 6, \ldots$$
$$= 0, \qquad\qquad n = 3, 5, 7, \ldots$$

The power spectrum is $P_n = |\alpha_n|^2$; therefore,

$$P_0 = \frac{1}{\pi^2}, \qquad P_1 = \frac{1}{16},$$

$$P_n = \frac{1}{\pi^2(n^2 - 1)^2}, \qquad n = 2, 4, 6, \ldots$$
$$= 0, \qquad\qquad n = 3, 5, 7, \ldots.$$

### C6.11

If $v(t)$ in the circuit shown is the halfwave-rectified cosine wave of Problem C6.4, find the coefficients of the exponential Fourier series of the current $i(t)$.

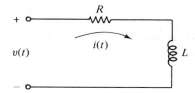

**Solution:**

The appropriate system function in this case is the admittance,

$$\frac{I}{V} = H(j\omega) = \frac{1}{R + j\omega L}$$

$$H(jn\omega_0) = \frac{1}{R + jn\omega_0 L} = \frac{1}{R + jnL}.$$

Using the $\alpha$ coefficients of $v(t)$ derived in Problem C6.7, we obtain for the $\alpha$ coefficients of the current $i(t)$:

$$\alpha_0 = \frac{1}{\pi R}, \qquad \alpha_1 = \frac{1}{4} \frac{1}{R + jL}$$

$$\alpha_n = \frac{(-1)^{(n/2)+1}}{\pi} \frac{1}{n^2 - 1} \frac{1}{R + jnL}, \qquad n = 2, 4, 6, \ldots$$
$$= 0, \qquad\qquad\qquad n = 3, 5, 7, \ldots.$$

### C6.15

Find the Fourier transform of the function

$$f(t) = \epsilon^t, \qquad -1 \le t \le +1$$
$$= 0, \qquad \text{elsewhere.}$$

*Solution:*

$$F(j\omega) = \int_{-\infty}^{+\infty} f(t)\epsilon^{-j\omega t}\, dt = \int_{-1}^{+1} \epsilon^t \epsilon^{-j\omega t}\, dt = \int_{-1}^{+1} \epsilon^{(1-j\omega)t}\, dt$$

$$= \frac{1}{1-j\omega}[\epsilon^{(1-j\omega)} - \epsilon^{-(1-j\omega)}] = \frac{2\sinh(1-j\omega)}{1-j\omega}.$$

### C6.16

Find an expression for the energy density spectrum of the function of Problem C6.15.

*Solution:*

From Problem C6.15,

$$F(j\omega) = \frac{\epsilon^{(1-j\omega)} - \epsilon^{-(1-j\omega)}}{1-j\omega}.$$

In order to find the magnitude squared of this function we will express the numerator in terms of its real and imaginary parts:

$$\epsilon^{(1-j\omega)} - \epsilon^{-(1-j\omega)} = \epsilon\epsilon^{-j\omega} - \epsilon^{-1}\epsilon^{j\omega} = \epsilon(\cos\omega - j\sin\omega)$$

$$- \epsilon^{-1}(\cos\omega + j\sin\omega) = (\epsilon - \epsilon^{-1})\cos\omega - j(\epsilon + \epsilon^{-1})\sin\omega$$

$$= 2.35\cos\omega - j3.086\sin\omega.$$

Therefore,

$$F(j\omega) = \frac{2.35\cos\omega - j3.086\sin\omega}{1-j\omega}$$

$$S(\omega) = \frac{1}{2\pi}|F(j\omega)|^2 = \frac{1}{2\pi}\frac{5.522\cos^2\omega + 9.523\sin^2\omega}{\omega^2 + 1}.$$

## CHAPTER 7

### C7.1, C7.2

Show that if $f(t) = (a/2)\epsilon^{-a|t|}$, then $\lim_{a\to\infty} f(t) = \delta(t)$.

*Solution:*

$f(0) = a/2$; $\lim_{a\to\infty} f(0) = \infty$. For any $t_1 \neq 0$, $\lim_{a\to\infty} f(t_1) = 0$. (Why?)

$$\int_{-\infty}^{+\infty} f(t)\, dt = 2\int_0^{\infty} f(t)\, dt = a\int_0^{\infty} \epsilon^{-at}\, dt = 1, \qquad \text{for all } a.$$

**C7.3**

Evaluate the following integral:

$$\int_0^\infty [\delta(t+1) + \delta(t-1)] f(t)\, dt.$$

*Solution:*

Only the delta function $\delta(t-1)$ is within the interval of integration; therefore,

$$\int_0^\infty \delta(t+1) f(t)\, dt + \int_0^\infty \delta(t-1) f(t)\, dt = 0 + f(1) = f(1).$$

**C7.4**

Find the current inpulse response of the circuit shown by evaluating the step response and differentiating:

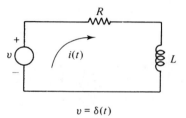

$$v = \delta(t)$$

*Solution:*

For a unit step function of applied voltage,

$$i(0) = 0, \qquad i(\infty) = \frac{1}{R}, \qquad \tau = \frac{L}{R}.$$

Therefore,

$$i(t) = \frac{1}{R}(1 - \epsilon^{-t/\tau})\, u(t).$$

If we differentiate this current, we obtain

$$\frac{di(t)}{dt} = \frac{-1}{R}\left(-\frac{R}{L}\right)\epsilon^{-t/\tau} u(t) = \frac{1}{L}\epsilon^{-t/\tau} u(t),$$

which is the result obtained in Example 7.1.

### C7.7

Sketch the result of convolving the pair of functions indicated.

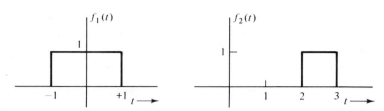

### Solution:

Let us turn around $f_2(\tau)$. At $t = 0$, $f_1(\tau)$ and $f_2(t - \tau)$ appear as shown:

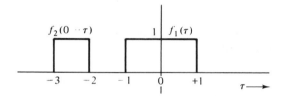

As $t$ increases from zero, we begin to get overlap of the functions at $t = 1$, with complete overlap for $2 \leq t \leq 3$. The convolved result is as shown.

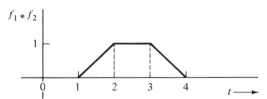

### C7.8

If $x(t) = \epsilon^{-t}u(t)$ and $y(t) = u(t) - u(t - 1)$, find $z(t) = x(t) * y(t)$.

### Solution:

We will solve this problem by (a) turning around $x(\tau)$ and (b) turning around $y(\tau)$.

(a) For $t \leq 0$, $z(t) = 0$.

For $0 \leq t \leq 1$, $z(t) = \int_0^t 1 \cdot \epsilon^{-(t-\tau)} \, d\tau = \epsilon^{-t} \int_0^t \epsilon^\tau \, d\tau = 1 - \epsilon^{-t}$.

For $t \geq 1$, $z(t) = \int_0^1 1 \cdot \epsilon^{-(t-\tau)} \, d\tau = \epsilon^{-t} \int_0^1 \epsilon^\tau \, d\tau = \epsilon^{-t}(\epsilon - 1)$

$\qquad = 1.7183\epsilon^{-t}.$

(b) For $t \leq 0$, $z(t) = 0$.

For $0 \leq t \leq 1$, $z(t) = \int_0^t 1 \cdot \epsilon^{-\tau} \, d\tau = 1 - \epsilon^{-t}$.

For $t \geq 1$, $z(t) = \int_{t-1}^t 1 \cdot \epsilon^{-\tau} \, d\tau = \epsilon^{-t+1} - \epsilon^{-t} = \epsilon^{-t}(\epsilon - 1)$
$$= 1.7183\epsilon^{-t}.$$

Thus we obtain the same result by either method (*a* or *b*).

## CHAPTER 8

### C8.1

Show that

$$\mathcal{L}[\sin \omega_0 t \, u(t)] = \frac{\omega_0}{s^2 + \omega_0^2}.$$

*Solution:*

$$\int_0^\infty \sin \omega_0 t \, \epsilon^{-st} \, dt = \frac{\epsilon^{-st}(-s \sin \omega_0 t - \omega_0 \cos \omega_0 t)}{s^2 + \omega_0^2} \Big|_0^\infty.$$

For $\sigma > 0$ (the abscissa of absolute convergence),

$$\epsilon^{-st} = \epsilon^{-\sigma t}(\cos \omega t - j \sin \omega t) \xrightarrow[t \to \infty]{} 0.$$

Therefore, putting in the lower limit, we obtain

$$\mathcal{L}[\sin \omega_0 t \, u(t)] = \frac{\omega_0}{s^2 + \omega_0^2}.$$

The same result may be obtained by noting that

$$\mathcal{L}[\epsilon^{\alpha t} \, u(t)] = \frac{1}{s - \alpha}.$$

Therefore,

$$\mathcal{L}[\sin \omega_0 t \, u(t)] = \left[\frac{1}{2j}(\epsilon^{j\omega_0 t} - \epsilon^{-j\omega_0 t})\right] = \frac{1}{2j}\left(\frac{1}{s - j\omega_0} - \frac{1}{s + j\omega_0}\right)$$

$$= \frac{\omega_0}{s^2 + \omega_0^2}.$$

### C8.4

The circuit shown is in the constant steady state at $t = 0-$. At $t = 0$, the switch is closed. Find $I_2(s)$ as the ratio of two determinants.

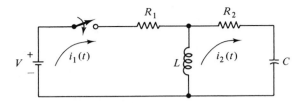

**Solution:**

The differential equations for $i_1$ and $i_2$ are

$$V - R_1 i_1 - L \frac{d}{dt}(i_1 - i_2) = 0$$

$$L \frac{d}{dt}(i_2 - i_1) + R_2 i_2 + \frac{1}{C} \int_{-\infty}^{t} i_2(\tau) \, d\tau = 0.$$

Since $i_L(0+) = v_C(0+) = 0$, the equations transform as follows:

$$(Ls + R_1)I_1 - LsI_2 = V$$

$$-LCs^2 I_1 + (LCs^2 + R_2Cs + 1)I_2 = 0.$$

Therefore, by Cramer's rule,

$$I_2(s) = \frac{\begin{vmatrix} Ls + R_1 & V \\ -LCs^2 & 0 \end{vmatrix}}{\begin{vmatrix} Ls + R_1 & -Ls \\ -LCs^2 & LCs^2 + R_2Cs + 1 \end{vmatrix}}.$$

## C8.13

Apply the initial- and final-value theorems to the transform

$$F(s) = \frac{s}{s^2 + \omega_0^2}.$$

**Solution:**

If we apply the initial-value theorem, we obtain

$$f(0) = \lim_{s \to \infty} sF(s) = \lim_{s \to \infty} \frac{s^2}{s^2 + \omega_0^2} = 1.$$

If we apply the final-value theorem, we obtain

$$f(\infty) = \lim_{s \to \infty} sF(s) = \lim_{s \to \infty} \frac{s^2}{s^2 + \omega_0^2} = 0.$$

However, since $f(t) = \cos \omega_0 t \, u(t)$, the result obtained by applying the final-value theorem is not valid, since $f(t)$ does not approach a limit as $t \longrightarrow \infty$; i.e., a final value does not exist!

### C8.20

Find all values of $s$ that satisfy the equation

$$\epsilon^s + 1 = 0.$$

**Solution:**

$$\epsilon^s = \epsilon^{\sigma + j\omega} = \epsilon^\sigma \epsilon^{j\omega} = -1.$$

The magnitudes and arguments of both sides of the equation must be equal, and

$$|-1| = 1, \qquad \underline{/-1} = \pm n\pi \quad (n \text{ odd}).$$

Therefore,

$$|\epsilon^s| = |\epsilon^\sigma||\epsilon^{j\omega}| = \epsilon^\sigma = 1, \qquad \sigma = \ln 1 = 0$$

$$\underline{/\epsilon^s} = \underline{/\epsilon^\sigma} + \underline{/\epsilon^{j\omega}} = 0 + \omega = \omega = \pm n\pi \qquad (n \text{ odd}).$$

Therefore, the values of $s$ which satisfy the original equation are

$$s = \pm jn\pi \qquad (n \text{ odd}).$$

## CHAPTER 9

### C9.1

Express the following differential equation in state-variable form:

$$\ddot{x} + a\dot{x} + bx = cy + d\dot{y}.$$

**Solution:**

If we define the state variables as

$$x_1 \triangleq x$$
$$x_2 \triangleq \dot{x},$$

then the equations are

$$\dot{x}_1 = x_2$$
$$\dot{x}_2 = -bx_1 - ax_2 + cy + d\dot{y}.$$

These equations are not state equations, however, because of the $\dot{y}$ term on the right; we must define the state variables in such a way that we get rid of this term. If we try

$$x_1 \triangleq x$$
$$x_2 \triangleq \dot{x} - ey,$$

then the equations are

$$\dot{x}_1 = x_2$$
$$\dot{x}_2 = -bx_1 - ax_2 + cy + d\dot{y} - e\ddot{y},$$

and if we let $e = d$, we have the valid state equations

$$\begin{bmatrix} \dot{x}_1 \\ \dot{x}_2 \end{bmatrix} = \begin{bmatrix} 0 & 1 \\ -b & -a \end{bmatrix} \begin{bmatrix} x_1 \\ x_2 \end{bmatrix} + \begin{bmatrix} 0 \\ C \end{bmatrix} y.$$

Hence $x_1 = x$ and $x_2 = \dot{x} - dy$ is a valid set of state variables for the original equation. (A general procedure for handling higher-order derivatives of the forcing function may be found in R. J. Schwarz and B. Friedland, *Linear Systems*, McGraw-Hill Book Company, New York, 1965.)

# *Index*

## DATE DUE

| OC 6 '78 | | | |
|---|---|---|---|
| NO 2 9 '78 | | | |
| DE 1 2 '78 | | | |
| NO 19 '80 | | | |
| | | | |
| | | | |
| | | | |
| | | | |
| | | | |
| | | | |
| | | | |
| | | | |
| | | | |
| | | | |
| | | | |
| | | | |
| | | | |

PRINTED IN U.S.A